Soft Error Reliability Using Virtual Platforms

Felipe Rocha da Rosa • Luciano Ost • Ricardo Reis

Soft Error Reliability Using Virtual Platforms

Early Evaluation of Multicore Systems

 Springer

Felipe Rocha da Rosa
Arm (United Kingdom)
Cambridge, UK

Luciano Ost
Loughborough University
Loughborough, UK

Ricardo Reis
Instituto de Informatica
Univ Federal do Rio Grande do Sul
Porto Alegre
Rio Grande do Sul, Brazil

Disclaimer: The views and opinions expressed in this book are those of the authors and do not necessarily reflect the official policy or position or Arm LTD.

ISBN 978-3-030-55706-5   ISBN 978-3-030-55704-1   (eBook)
https://doi.org/10.1007/978-3-030-55704-1

This Springer imprint is published by the registered company Springer Nature Switzerland AG
The registered company address is: Gewerbestrasse 11, 6330 Cham, Switzerland

Preface

The increasing computing capacity of multicore components such as processors and GPUs offers new opportunities for embedded and high-performance computing (HPC) domains. The progressively growing computing capacity of multicore-based systems enables to efficiently perform complex application workloads at a lower power consumption compared to traditional single-core solutions. Such efficiency and the ever-increasing complexity of application workloads encourage the industry to integrate more and more computing components into the same system. The number of computing components employed in large-scale HPC systems already exceeds a million cores, while 1000-cores on-chip platforms are available in the embedded community.

Beyond the massive number of cores, the increasing computing capacity, as well as the number of internal memory cells (e.g., registers and internal memory) inherent to emerging processor architectures, is making large-scale systems more vulnerable to both hard and soft errors. Moreover, to meet emerging performance and power requirements, the underlying processors usually run in aggressive clock frequencies and multiple voltage domains, increasing their susceptibility to soft errors, such as the ones caused by radiation effects. The occurrence of soft errors or single event effects (SEEs) may cause critical failures in system behavior, which may lead to financial or human life losses. While a rate of 280 soft errors per day has been observed during the flight of a spacecraft, electronic computing systems working at ground level are expected to experience at least one soft error per day in the near future. The increased susceptibility of multicore systems to SEEs necessarily calls for novel cost-effective tools to assess the soft error resilience of underlying multicore components with complex software stacks (operating system (OS) and drivers) early in the design phase.

The primary goal addressed by this book is the description of a novel fault injection framework, developed on the basis of the state-of-the-art virtual platforms, which integrates a set of novel fault injection techniques that enable to access the soft error reliability of complex software stack in a reasonable time. The proposed framework is extensively validated, though more than a million of simulation hours. The second goal of this book is to set the foundations for a new discipline in soft error reliability management for emerging multi/manycore systems using machine learning techniques. It will identify and propose techniques that can be used to provide different levels of reliability on the application workload and criticality.

Cambridge, UK   Felipe Rocha da Rosa

Loughborough, UK   Luciano Ost

Porto Alegre, Brazil   Ricardo Reis

May 2020

Contents

Acronyms

ASIC	Application-specific integrated circuit
COTS	Commercial off-the-shelf
DSE	Design space exploration
DBT	Dynamic binary translation
DRAM	Dynamic random-access memory
DVFS	Dynamic voltage and frequency scaling
FIM	Fault injection module
GPU	Graphics processing unit
HPC	High-performance computing
ISA	Instruction set architecture
JIT	Just-in-time
MPI	Message passing interface
MIPS	Million instructions per second
MBU	Multiple bit upset
MET	Multiple event transient
nm	Nanometer
OVP	Open virtual platforms
OVPsim	Open virtual platforms simulator
OS	Operating system
PE	Processing units
SDC	Silent data corruption
SMT	Simultaneous multithreading
SEU	Single event upset
SER	Soft error rate
SMP	Symmetric multi-processor
KIPS	Thousand instructions per second
TMR	Triple modular redundancy
VA	Virtual address
VP	Virtual platform
VP-FIM	Virtual platform fault injection module

Chapter 1
Introduction

Computers become ubiquitous in our modern society ranging from everyday life devices (e.g., televisions, vending machines, smartphones) to complex systems such as those used to weather forecast or search for microscopic subatomic particles [23, 57]. The continuous technology scaling [20] and the advance of multicore components such as processors and graphics processing unit (GPUs) are driving the microelectronics industry forward. This evolution can be exemplified by the advanced driver-assistance systems (ADAS),[19, 59, 70, 105] enabling self-driven cars in the near future [39]. The emerge of *The Internet of Things* (IoT) is another example[83, 112] which is expected to integrate about 30 billion of devices by 2020 [77]. The semiconductor systems dissemination phenomenon was possible (or caused) by Gordon Moore's seminal work [73] on transistor scaling. This work introduced the famous Moore's Law, which states that the number of transistors per square inch doubles every 18 months. As a consequence, every new technology node had delivered increasing performance, lower power consumption, and smaller transistor cost. The top plot of Fig. 1.1 displays the increasing number of transistors in microprocessors since 1970.

Single-thread processors have benefited from technological advancements to improve their performance by increasing clock frequency [20]. However, in mid-2000, this trend reaches the physical limits due to the increased power consumption and the current density within the chip [36, 37]. The central plot of Fig. 1.1 shows the growing number of cores in commercial processors over the last decades. Integrating modern multicore processors (e.g., big.LITTLE [4]) and GPUs in the same system is now commonplace in both embedded and high-performance computing (HPC) domains [20, 29]. Such systems aim to perform complex software stacks (i.e., operating system OS, drivers, and applications) from diverse fields (e.g., spatial, avionics, automotive). However, the ever-increasing demand for performance, energy efficiency, and high reliability of emerging systems is imposing a myriad of challenges to the design of underlying systems:

F. Rocha da Rosa et al., *Soft Error Reliability Using Virtual Platforms*, https://doi.org/10.1007/978-3-030-55704-1_1

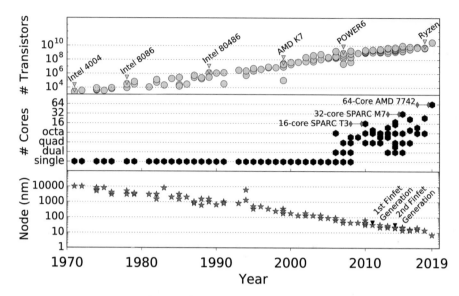

Fig. 1.1 Evolution of commercial processors during the last decades considering the number of transistors (top), number of cores (middle), and associated technology node (bottom) from 1970 to 2019. Information gathered from multiple sources including the ITRS (https://www.itrsgroup.com/) and https://en.wikipedia.org/wiki/Transistor_count

- *programmability*, ease of programming is a feature of paramount importance in large-scale systems composed of different processors, resulting in different platform libraries (e.g., MPI, OpenMP), compilers, instruction set architecture (ISAs) [38].
- *security*, with the increasing number of components and devices sending and receiving user sensitive data, providing a secure service is fundamental. The increasing system complexity introduces vulnerabilities to software and hardware architectures compromising the system behavior. Attackers can exploit such vulnerabilities to introduce malicious code, which may incur in undesired effects, security breaches, or damage to a system [70].
- *energy efficiency*, while the constant supply and threshold voltage scaling in transistors led to an exponential increase of leakage current. In addition to that, other physical restrictions related to device packaging, intra-chip current distribution, and cooler dissipation during power peaks further impact on the systems energy consumption [113]. Energy efficiency is becoming more critical than high-speed operation, and dark silicon era is imposing more power-oriented constraints to the design of such systems.
- *reliability and dependability*, the technology transistors reach the operation physical limits, thus becomes increasingly difficult for the hardware components to achieve reliable execution. The unreliability of multicore-based systems is emerging from several sources, e.g., electrical noise, cross-talk, radiation particles, aging, and variability [63].

Reliability is rapidly emerging as a significant design metric in both embedded and HPC domains. The increasing chip power densities allied to the continuous technology shrink is making emerging multicore-based systems more vulnerable to soft errors, such as the ones caused by radiation events [63]. As illustrated in the bottom graph of Fig. 1.1, commercial processors based on 10 nm process node are likely to be available in the market in the coming years. Until recently, radiation-induced faults issue was relegated to high-availability systems such as military applications and radiation-bounded systems as spatial and avionics. Nowadays, the occurrence of soft error appears as a primary concern of electronics systems working at ground level [74]. A transient error, also known as a soft error, induced by radiation particles can lead to financial or human life losses [110]. For instance, the occurrence of a soft error in HPC systems could lead to the underutilization of resources, which results in extra cost and time wasted waiting for the re-execution of applications/jobs. Although a supercomputer with 900 compute nodes registered a rate of 0.15 error per day, over a year of operation [11], electronic computing systems working at ground level are expected to experience at least one soft error per day in near future [43].

The resulting growing susceptibility of multicore systems to soft errors necessarily calls for novel cost-effective tools to assess the soft error resilience of underlying systems early in the design phase. The preceding context provides the *motivation* for this book, which aims at investigating novel fault injection techniques and tools that can be used to assess soft errors of multicore-based systems under fault campaigns at early design time.

1.1 Hypothesis to Be Demonstrated in This Book

This book relies on two hypothesis:

- Enhancing virtual platforms with fault injection capabilities enable to speed up the evaluation of more realistic multicore systems (i.e., real software stacks, state-of-the-art ISAs) at early design phases. The use of such fault injection frameworks increases the probability of generating soft errors and failures in those multicore systems, which allows to generate and collect more error/failure-related data in a reasonable time.
- With large error/failure-related data sets, engineers are more likely to identify meaningful relationships or associations between fault injection results, application characteristics, and platform parameters. When dealing with large failure-related data sets obtained from the fault campaigns, it is essential to filter out non-correlated features (i.e., Parameters without a direct relationship with the system reliability). In this regard, the second hypothesis of this book is that the use of supervised and unsupervised machine learning techniques is appropriate to filter and identify the correlation between fault injection results and application and platform characteristics, enabling them to improve existing fault mitigation techniques, as well as to investigate and propose new and more efficient ones.

1.2   Book Goal

In order to address the hypothesis mentioned above, the strategic goal of this book is first to combine existing and novel simulation and fault injection techniques into virtual platforms, targeting fast and detailed soft error reliability exploration of state-of-the-art multicore systems. The second strategic goal of this book is investigate appropriate machine learning techniques to develop an automated engine capable of searching and identifying the individual (or combinations of) microarchitectural (e.g., instruction types, memory stats) and software parameters (e.g., number of branches, etc.), which present the most substantial relation relationship with each detected soft error and failures.

To accomplish the first strategic goal, the following specific objectives were considered and are further explained in the next chapters:

- Identification of the most suitable and efficient virtual platforms to include fault injection capabilities, aiming to support the soft error analysis of state-of-the-art processor models;
- Analysis and port of several benchmarks from embedded and HPC domains, including the Rodinia and NASA NAS Parallel Benchmark (NPB) suites;
- Investigation of the soft error analysis consistency between an instruction-accurate VP and a cycle-accurate full system simulator;
- Proposal and development of novel fault injection techniques and tools that enable to trace, evaluate, and identify particular source of errors (e.g., application functions, code structure).

To achieve the second goal, the following tasks were identified and their realization are detailed covered in Chap. 6:

- Exploration of machine learning techniques that can be used to enable the identification of individual (or combinations of) microarchitectural and software parameters that present the most substantial relation relationship with each detected soft error or failure.
- Evaluate the benefits and drawbacks of proposed techniques and developed tools when investigating the impact of software and hardware parameters on soft error system resiliency, considering a significant number of fault injection campaigns.

1.3   Original Contributions of This Work

Figure 1.2 illustrates the main contributions of this work, which are joined into a soft error analysis flow that includes fault injection and soft error analysis extensions (fully described in Chap. 3), and automated soft error correlation using machine learning techniques (described in Chap. 6).

The main contributions of this book are described as follows:

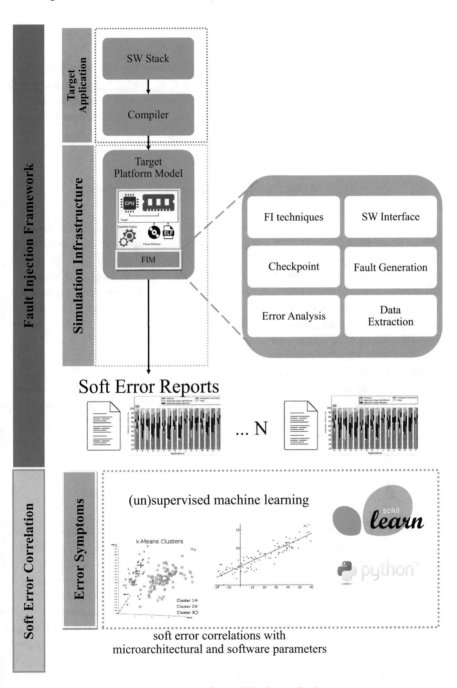

Fig. 1.2 Proposed design space exploration flow and book contributions

1.3.1 Early Soft Error Evaluation

Proposal of two flexible fault injection (FI) frameworks: the OVPsim-FIM developed on top of the OVPsim [56] and the gem5-FIM, which relies on the gem5 [18]. Both frameworks integrate a set of fault injection techniques, allowing fast soft error susceptibility exploration considering state-of-the-art multicore processor architectures, such as ARMv7, ARMv8, and big.LITTLE.

1.3.2 Novel Non-intrusive Fault Injection Techniques

Random fault injection homogeneously probes the application (i.e., every function has an equal fault probability), nevertheless, some code segments are more critical than others to the system reliability. The OVPsim-FIM was extended with four novel and non-intrusive FI techniques targeting: (1) the virtual memory, (2) variables, (3) function code, and (4) function execution. The underlying techniques offer flexibility and full control over the fault injection process when targeting complex software stacks. Further, this new module provides a more powerful fault inspection module, enabling the user to add custom verifications. In other words, it is possible to check data structures and execution patterns during or after the application simulation. For example, this new module reduces false-positive SDC (i.e., silent data corruption) detections by checking only a select group of variables instead of the entire memory. This tool provides software engineers with detailed and comprehensive fault injection capabilities to explore critical code sections in a quick and non-intrusive manner.

1.3.3 Instruction-Accurate Fault Injection Consistency

Instruction-accurate simulators provide a simulation performance of thousands of millions of instructions per second (MIPS), enabling quick explorations of large and complex scenarios. However, its accuracy regarding soft error assessment was never addressed and to investigate this matter, the proposed OVPsim-FIM (i.e., instruction-accurate simulator) was compared against the cycle-accurate gem5-FIM. This exploration comprises millions of fault injections. Results show that the unmodified OVPsim-FIM can achieve an average mismatch of less 25% when considering the gem5-FIM. This work further explores the OVPsim-FIM engine by investigating different configurations. With the proper simulator settings, the average mismatch can be reduced to less than 15%. More interesting, the worst case mismatch can be reduced by fivefold while sustaining the high-performance simulation suitable to early design space explorations.

1.3.4   Extensive Investigation of the Software Stack Impact on the System Reliability

This work uses the proposed FI framework scalability to explore early design decisions impact on the system reliability, e.g., architecture, number of cores, ISA, OS, parallelization library, among other possible configurations. Different from other works, the promoted frameworks use a realistic software stack comprising multiple unmodified operating systems (e.g., Linux 4.3, Linux 3.13, FreeRTOS) alongside parallelization libraries (e.g., OpenMP, MPI, PTHREADS, and OmpSs). Further, this work addresses another common issue of fault injection frameworks: *performance* and *scalability*. The developed fault injector adopt three simulation techniques to increases the soft error analysis performance: (1) host multicore parallelization, (2) checkpoint and restore technique, and (3) large-scale and distributed simulation, targeting its use on supercomputers.

This extensive evaluation considers more than three million fault injections (requiring up to three million of simulation hours) targeting 45 distinct benchmarks, considering serial, MPI, and OpenMP implementations from the NAS Parallel Benchmark (NPB) suite [8] and the Rodinia benchmark suite [25] among others. This exploration targets single, dual, quad, and octa-core ARM Cortex-A9 and ARM Cortex-A72 processor models. The investigation shows distinct effects of the chosen parallelization (e.g., OpenMP vs. MPI) library on the system fault tolerance, also, how the ISA decision can impact the system behavior under fault influence.

1.3.5   Correlating Soft Errors and Microarchitectural Data

Converting fault injection explorations into actual system reliability improvements is not a straightforward process. This work proposes a cross-layer investigation toolset that uses machine learning techniques to perform multivariable and statistical analysis using the gem5 microarchitectural information (e.g., memory usage, application instruction composition) along with other software profiling tools (e.g., line coverage) that are combined with soft error vulnerability evaluation results (i.e., fault injection campaigns). Proposed toolset enables to reduce the number of fault injection campaigns required during early design space exploration by using software symptoms (e.g., execution time, number of branches) correlated with soft error vulnerabilities to improve the target application reliability. Developed toolset provides users with a flexible investigation, where several information sources can be easily included, selected, and conformed to different machine learning investigation techniques.

1.4   Book Outline

This book is organized into six chapters where the previous chapter introduced the reliability issues in modern system design and this book contributions for this space. The following paragraphs present a short book summary chapter by chapter.

Chapter 2: Background on Soft Errors This chapter introduces the modern challenges to modern electronic devices in Sect. 2.1, then Sect. 2.2 presents a brief background on soft errors history and source mechanisms. While Sect. 2.3 presents the state of the art on soft error assessment using distinct approaches.

Chapter 3: Simulation-Based Fault Injection Using Virtual Platforms First, Sects. 3.1 and 3.2 provide a discussion on the available virtual platforms and their fault injection frameworks. Section 3.3 describes the adopted fault model, while Sect. 3.4 presents the fault injection flow. The fault injection frameworks using the gem5 and OVPsim simulators are detailed by Sects. 3.5 and 3.6. Section 3.7 provides extra tooling to help software developers to explore in profound the application effects under soft errors.

Chapter 4: Performance and Accuracy Assessment of Fault Injection Frameworks Based on VPs In this chapter, the authors evaluate the fault injection consistency of fault injection frameworks using virtual platforms. Section 4.1 describes the 45 adopted applications from the Rodinia benchmarks and the NAS Parallel Benchmark. Then, Sect. 4.2, explores the performance and accuracy of the proposed OVPsim-FIM.

Chapter 5: Extensive Soft Error Evaluation Next, the authors explore the flexibility of virtual platforms to perform complex soft error reliability investigations. Section 5.1 evaluates soft error considering multicore design decisions (e.g., ISA). Focused and detailed fault injection are performed in Sect. 5.2.

Chapter 6: Machine Learning Applied to Soft Error Assessment in Multicore Systems This chapter describes the promoted cross-layer investigation tool which performs multivariable and statistical analyses. First, we introduce some basic machine learning concepts in Sect. 6.1. Section 6.2 debates the state of the art of reliability using machine learning techniques. Sections 6.3 and 6.4 discuss, respectively, the problem of investigating large-scale fault injection campaigns and how this work mitigates this issue. The proposed tool requires multiple machine learning techniques which are described by Sect. 6.5, while Sect. 6.6 details the tool execution flow. Finally, Sect. 6.7 shows the results related to multiple investigations using this proposed tool.

Chapter 2
Background on Soft Errors

This chapter details some necessary backgrounds and state-of-the-art works related to this work exploration on soft error assessment. First, Sect. 2.1 enumerates several reliability challenges encountered by modern electronic devices, while Sect. 2.2 provides a brief history and background on radiation-induced soft errors. In particular, Sect. 2.2.1 explains the particle strike and the charge accumulation process, Sect. 2.2.2 shows several fault masking mechanisms, and Sect. 2.2.3 introduces useful metrics to evaluate the occurrence of soft errors. Section 2.3 investigates the recent innovations in soft error vulnerability assessment.

2.1 Main Reliability Challenges in Electronic-Based Systems

Semiconductor industry is facing significant reliability challenges to guarantee the correct functionally of electronic systems [63]. Problems such as process variability, permanent faults, and transient faults are significant issues for semiconductor-related industry sectors [51, 97]. Three main fault groups comprehensive encompass the device reliability challenges for future technologies:

1. **Process Variability**: As the transistor's features scale down the chip variability grows during the fabrication process [80] impacting multiple parameters, e.g., channel length, doping concentration, oxide thickness. For comparison's sake, the state-of-the-art fabrication process (e.g., 10nm) shapes are only four times larger 20 than the diameter of a DNA strand. Consequently, devices with identical logical design have distinct physical and electrical characteristics from die to die (i.e., inter-die variations) and on the same die (i.e., intra-die variations). These issues result in yield, power, and performance reduction as the same design must withstand a wider parameter variation [21].

© The Editor(s) (if applicable) and The Author(s), under exclusive license to
Springer Nature Switzerland AG 2020
F. Rocha da Rosa et al., *Soft Error Reliability Using Virtual Platforms*,
https://doi.org/10.1007/978-3-030-55704-1_2

2. **Permanent faults** are physical imperfections such as stuck-at-zero and timing violations in conjunction with aging problems. For instance, decreasing transistor sizes cause faster aging and eventually transistor wear out due to distinct phenomena such as Hot Carrier Injection (HCI), Bias Temperature Instability (BTI), Electromigration, and Time-Dependent Dielectric Breakdown (TDDB) [2]. These problems reduce the chip average lifetime and impact on its meantime between failures (MTBF) by steadily increasing propagation delays.

3. **Transient faults** or soft errors encompass all sort of malfunctions without permanent circuit damage. Such errors may occur due to the occurrence of electrical noise, electromagnetic interference, as well as exposition to radiation. Soft errors cause single event effects (SEEs) in a processor, which can be propagated through logical (e.g., logic gates) and memory elements (e.g., latches, registers). Whenever an SEE surpasses a specific charge threshold[1] it will induce an incorrect computation affecting logical and storage elements. In contrast to permanent faults and process variability, new data can still be correctly written and stored on the affected device. Due to the technology high-frequency, low voltage supply, and continuous technology shrink the transistor becomes more susceptible to soft errors as the minimal energy to provoke it is reducing [10, 93].

Process variability increases the production cost and development time, while permanent faults lead to premature wearing (i.e., reducing the system's lifetime). However, in most of the case, the correct fabrication process and new low-level system design (i.e., gate-level) can partially mitigate its occurrence. Also, its effects appear during a prolonged period enabling early detection and correction. In contrast, transient faults will introduce erroneous behavior in the system at random time without previous warning or at predictable rates. Soft errors may emerge either during intensive or idle working periods, affecting both critical and non-critical system functionalities. Among the aforementioned reliability challenges, soft errors are the most prominent one in several sectors of the semiconductor industry. Processor-based systems working at sea level are expected to experience at least one soft error per day [43].

2.2 Radiation-Induced Soft Errors

In recent years, soft errors caused by radiation particles became a recurring research topic in both academia and industry [66, 98]. Nevertheless, the radiation effects on semiconductor materials are well-known for more than half-century. In 1961, [106] predicted the transistor scaling limits around 10 μm for several reasons, including cosmic rays. A decade later in 1975, the misbehavior of digital circuits used in a communication satellite was identified and studied by Binder et al. [17].

---

[1]This critical threshold depends on the technology node, cell design, and neighboring gates.

Binder et al. reported galactic cosmic rays effects in the processor flip-flops as the principal error source. In a retrospective investigation, [78] found an unusual number of parity errors in the Los Alamos Cray-1 supercomputer during the year of 1976, pointing to ground-level high-energy neutrons as its cause. DRAM soft error vulnerability to heavily-ionizing radiation are discussed by May and Woods [69] and Ziegler and Lanford [114] investigate the silicon interactions at sea level with cosmic-ray nucleons and muons. The next subsection presents examples of radiation strike mechanisms. Afterwards, Sect. 2.2.2 discusses masking and propagation mechanisms, Sect. 2.2.3 shows some soft errors, quantification metrics, and Sect. 2.2.4 explores the trends of soft errors in future systems.

2.2.1 Radiation Source and Soft Errors Mechanisms

According to Baumann et al. [10] soft errors induced by radiation originate from three primary sources: (1) the emissions of **alpha particles** due to the presence of radioactive impurities on the chip packaging; (2) an alpha particle (i.e., two neutrons and two protons) traveling through a semiconductor material loses kinetic energy leaving an ionization trail behind; and (3) **high-energy cosmic rays** originating from outer space, which produce a complex cascade of secondary particles in earth's atmosphere (e.g., muons, protons, neutrons, and pions). For example, a neutron collision with one Si atom emits lighter ions and sub-particles (e.g., protons, alpha particles). **Low-energy cosmic rays** (i.e., up to 1 MeV) create ionizing particles in electronic devices from the interaction of neutrons and borons atoms (i.e., p-type dopant).

The collision of a sub-atomic particle induces a single event transient (SET) by generating secondary particles capable of ionizing the n-p junctions of sensitive transistors causing a voltage charge or discharge in the stroke node. Modern and smaller technology can also suffer from multiple event transients (MET) as the same particle induces a charge in several neighboring transistors. For simplicity's sake, this subsection focuses on single event effects. Figure 2.1a–c shows a high-energy ion path through a reverse-biased junction, i.e., the most charge-sensitive circuit node [10, 66]. Figure 2.2 displays the corresponding current pulse resulting from the following three phases:

1. The high-energy particle interaction with the silicon transfers kinetic energy to the semiconductor material creating a track of electron–hole pairs and forming a conical high carrier concentration in the wake of the energetic ions passage (Fig. 2.1a).
2. The electric field in the depletion region collects the closest charge carriers creating a transient current/voltage at the target device node (Fig. 2.1b). During this phenomenon, a temporary channel may be formed for a short period, around few picoseconds.

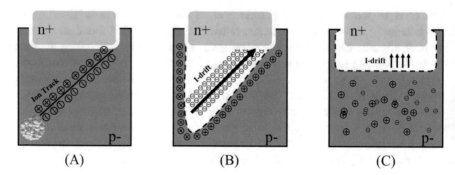

Fig. 2.1 Charge generation and collection phases in a reverse-biased junction

Fig. 2.2 Resultant current
pulse caused by the passage
of a high-energy ion in a
reverse-biased junction

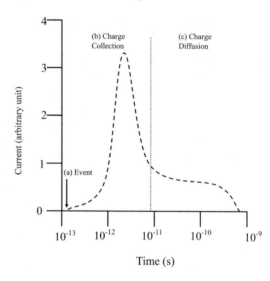

3. A nanosecond later the diffusion begins to dominate the collection process
   Fig. 2.1c, inducing additional current formation.

The radiation event deposits a certain amount of charge (Q_{all}) due to the hole
pair formation [53, 55]. The collection mechanism described above has a maximum
efficiency coefficient (L_{max}) and depends on the ion track length (x_c). For instance,
the total amount of collected charge by a SRAM is giving by Eq. (2.1) [55]:

$$Q_{collected} = Q_{all} \left(\frac{x_c}{L_{max}} \right) \tag{2.1}$$

$Q_{collected}$ values range from one to several hundred pC, and the precise $Q_{collected}$
estimation involves the ion strike angle, path, energy, mass, and point of impact
considering the nearest reverse-biased junction. The device geometry and electrical
characteristics have additional influence on the collection process. The displacement

of charge carriers in the time creates an electrical current in the target devices modeled by Eq. (2.2):

$$I(t) = \frac{Q_{collected}}{\tau_\alpha - \tau_\beta} \left(e^{-\frac{t}{\tau_\alpha}} - e^{-\frac{t}{\tau_\beta}} \right) \qquad (2.2)$$

τ_α and τ_β are process dependent constants denoting the collection time and ion track establishment time. The literature reports a typical value of 164 ps for the τ_α and 50 ps for the τ_β [66, 79]. The charge collection in the stroke node may lead to a single event upset (SEU), in other words, introducing an incorrect bit in the memory cell. The quantity of $Q_{collected}$ required to create an SEU in a device is denoted by the critical charge ($Q_{critical}$) and expressed by:

$$Q_{critical} = \int_0^{T_F} I_D(t)\, dt \qquad (2.3)$$

In Eq. (2.3), $I_D(t)$ represents the time-dependent drain transient current, and [79] defines the flipping time (T_F) as the time instant when the struck transistor drain voltage is equal to the gate voltage after the radiation event. In simple circuits such as DRAMs, an error only occurs whenever the $Q_{critical}$ is greater than $Q_{collected}$ masking otherwise the radiation event. The SRAM feedback loop can restore the original value if the recovery time[2] does not exceed the feedback time[3] [32].

2.2.2   Fault Propagation and Masking

The transient errors create unintentional electrical signals that need to travel through many design abstraction layers from the transistor until reaching the application variables and control flow. Several mechanisms can mask this fault: First, a particle strike needs to generate enough charge collection to create a noticeable electrical signal, which depends on the transistor electrical characteristics. The circuit resistance also attenuates the signal strength (i.e., electrical masking). The internal logic structure leads to *logical masking* when the fault propagation path is blocked by another dominating data path. The circuit timing requirements (e.g., setup and hold times) constrain all electrical signals, including faults. Whenever the faulty signal violates one delay constraints (i.e., the signal arrives either too early or too late to be captured during the clock edge) results in a *temporal masking*.

Digital systems present two main circuit components: Combinational (e.g., AND, OR, XOR) and sequential circuits (e.g., SRAM cells, latches, flip-flops). Combinational circuits are regarded as less prone to soft errors due to the above-

---

[2]Time taken for the struck node voltage to return to its pre-strike value.
[3]The time taken for the struck node voltage to become latched as incorrect data.

described masking mechanisms and the absence of feedback loop in the underlying circuit. In turn, the sequential circuits are more vulnerable to radiation events as a single strike has enough energy to reverse the stored data as a result of the feedback loop. Also, memory elements are susceptible to bit-flips during extended periods of time when holding data, in contrast, the sequential circuit switches (i.e., changes the value) more often. For example, an SRAM cell may be affected by SEUs during almost the entire clock cycle [93]. Also, the clock network under the influence of soft errors results in memory elements incorrect operation.

At *architecture level* (e.g., program counter, pipeline registers, register file, arithmetic, and logic unit), an erroneous bit can be further masked due to the write and the read operations. For example, a fault present in a register can be overwritten by a write before a read, and thus eliminate the incorrect bit. An error is a fault that propagated inside the application (or OS) before being perceived by the user. At the software level, SEUs are incorrect variable values or wrong control flow executions. An algorithm can overwrite the variable before its value is consumed masking the fault. Even with all those masking processes, the incidence of soft errors is increasing due to the technology susceptibility to radiation events. Well defined metrics and methods are necessary to analyze the impact of transient faults under different conditions considering system architecture, application, and compiler.

2.2.3 Soft Error Metrics

This subsection describes some useful soft error metrics: The transient errors occurrence per time is quantified by the Soft Error Rate (SER) and measured by the Failure-In-Time (FIT) parameter. The FIT is equivalent to the number of failures (i.e., soft errors in this context) in one billion of operation hours [10].

A well-established metric is the architectural vulnerability factor (AVF) [75] that estimates a particular bit susceptibility to create a visible error in the application. The architectural bits can be divided into two subgroups: (1) The bits required for an *architecturally correct execution* (ACE) and (2) the un-ACE bits. While an ACE bit propagates faults to the final output, un-ACE bit does not create a visible error. AVF use enables the search for the most vulnerable architectural state bits. The AVF focuses on the correct result, and thus bit fault can change the intermediary computations and still have an AVF of 0%. For instance, a transient fault in a branch predictor generates a miss-prediction, which potentially requires the re-execution of some instructions without altering the final application.

It lacks an explicit masking rate model varying according to each target hardware component. Consequently, this metric may lead to an over-estimation the number of errors. The register vulnerability factor (RVF) [109] is a metric explicitly used to measure the register file susceptibility to soft errors. The RVF accounts for the vulnerability factor by calculating the intervals between two vulnerable register operations (i.e., write/read and read/read). Write operations will mask any fault previously propagated, and thus, the interval between writes is the most vulnerable period for any application.

The AVF shows a larger granularity than desired to measure the instructions interaction. To address this issue, [22] create an instruction-based criticality assessment metric called Instruction Vulnerability Factor (IVF). This metric uses distinct fault injection techniques to probe every instruction in the code. Given the involved complexity, complete coverage can be difficult to achieve. Another instruction-oriented approach is the Instruction Vulnerability Index (IVI) [85]. Its composition includes the ACE bits, the area, and specific pipeline components vulnerability. This analytical approach avoids the exhaustive simulation by using a fault probability in each pipeline component. Also, the IVI metric allows the extension to register file IVI by incorporating the vulnerability window of each register in the instruction.[4]

2.2.4   Soft Error Trends in Electronic Systems

In the last decades, Moore's and the Dennard's laws coupled transistor features and V_{dd} scaling ratio. Simultaneously, the V_{dd} directly impacts on the memory cells sensitivity to radiation events due to the charge reduction to achieve the $Q_{critical}$. On larger designs as flip-flops[5] the area reduction plays a significant role to improve the cell reliability as V_{dd} reduction impacts on SRAM cells reliability. The scaling process reduces the cell sensible areas, consequently, the total collected charge [31, 35]. Nevertheless, the transistor density creates a new phenomenon: a single particle strike can induce charge generation in several neighboring cells (i.e., MET—multiple event transients), which may result in Multiple-Bit Upset (MBU). Also, MET events in sequential logic reduce the electrical masking probability due to the multiple concurrent propagations paths.

The difficulties in downscaling the planar CMOS technology lead the major semiconductor foundries, such as Intel, TSMC, and GlobalFoundries, to adopt the multi-gate nonplanar transistor technology also known as Fin Field-Effect Transistor (FinFET) [33, 90]. Radiation effects as SETs and SEUs have been studied and analyzed in bulk CMOS [53], and more recent FinFET has been proven equally susceptible to neutron and alpha particles strikes [5, 54].

The FinFET technology improves the system soft error reliability by reducing the gate width in favor of the height forming a tri-gate shape [94]. The tri-gate architecture reduces the drain/source area (i.e., reverse-biased junctions), and by consequence, diminishing the charge collection process. This phenomenon reduces the total collected charge and the critical charge[6] as shown by Seifert et al. Seifert

---

[4]The vulnerability window is the time between a register write and the last read in the value. During this period, the register is susceptible to propagate soft errors.

[5]When compared with SRAMs.

[6]Amount of induced charge in an SRAM node to change it data status.

et al. [94] report a 23 times SER improvement on a second generation 14-nm SRAM cell when compared to a 32-nm planar technology at nominal voltage. The amelioration from the first FinFET generation (i.e., 22-nm) to the second (i.e., 14-nm) achieves up to 8 times for the same cell. The newer generation is taller and slimmer than the previous one, and so further reducing the device sensitive areas.

The ITRS [60] reports soft errors as one major design issues to the sub-22 nm nodes. Although FinFET-based systems individual cell reliability has improved, the trend points to more vulnerable architectures [66, 94] as the amount of integrated memory grows at each new architecture (e.g., larger caches and more complex pipelines). Further, most works on circuit reliability consider only the system execution at nominal voltage and temperature, which does not reflect the reality and can affect directly the FinFET reliability [87, 88]. The high-energy neutrons are becoming the primary source of soft error at ground-level, with up to 77%, surpassing the alpha particles. To ensure the system's reliability or at least fail-safe functionality, the designer should be able to identify soft errors during the initial design cycle. Aiming at accelerating the fault injection evaluation at early design phases, researchers are investigating fast, and efficient simulation-based fault injection approaches to enable complex soft error resilience analysis regarding different system configurations at an acceptable time.

2.3 Soft Error Assessment

This section resumes the most widely adopted approaches to assess soft error effects on embedded systems. The development of techniques for soft error injection in processors has been studied to evaluate processor architecture, organization, and applications running on those processors in early product design phase. Fault injection can be performed on board-oriented approaches by interrupting the processors and forcing the processor to execute corrupted data that is modeling the fault. However, this approach is also very time-consuming when considering complex applications executing under millions of fault injection experiments. Another alternative is exposing the board to neutron radiation. Although radiation tests produce the most accurate results, they have a very high cost. Especially when looking for a large number of error events under neutron fluence, each test may easily take several days to build a desirable confident statistic.

The first group relies on detailed fault injection estimation at register or gate-level employing commercial tools, such as ModelSim from Mentor. For instance, a tool called VFIT integrates a series of VHDL-based elements designed around the ModelSim to support fault injection evaluation [9]. Authors considered 3000 faults per experiment. A similar work extends Modelsim capabilities in order to inject faults in RTL descriptions using Perl and Tcl scripts [89]. Authors in [103] use Foreign Language Interface (FLI) available in ModelSim to diminish control overheads during simulation. FLI enables C-Written modules to monitor and modify any signal in simulation time. Results include a compiled CORDIC processor in a

10,000 faults scenario reaching 2.33 faults per second in the best simulation case. Due to the number of modeled aspects [9, 89, 103], the detailed evaluation produces accurate soft error results although it is time-consuming, and the amount of memory required by these approaches is too high. Consequentially, the work restricts the experiments to few thousands of faults (e.g. [9, 103]) considering a single target processor or specific ISA. Other drawbacks of such approaches are the poor fault access, and design modifications are usually highly intrusive, which increase the design space exploration cost.

To speed up the fault injection simulation while improving modeling capabilities the second group emphasizes the use of SystemC [84]. The work proposed in [34] explores how faults can be modeled and injected at different SystemC abstraction levels (e.g., RTL, TLM). The authors in [14] focus on soft error evaluation and a NoC model was used as case study. In [95], a behavioral SystemC description of an MPEG-2 is used to validate a fault injection technique developed on the SystemC API basis. Authors in [72] coupled VHDL and SystemC in a hierarchical design fault simulation process based on SystemC, which can be employed at different abstraction levels. A Python-based framework called ResSP is proposed in [13, 15]. ResSP provides wrappers to assemble SystemC components and processor models (e.g., PowerPC, Leon2, and ARM7) described in ArchC [86]. Experiments comprise 10,800 faults injections in a Leon processor model connected to a memory through a bus. As reported in [15], ReSP simulation achieves 2–3 MIPS on a 2-core host. Although SystemC reduces the simulation cost, the lack of processor/ISA models and the still inadequate performance limit its adoption when exploring extensive fault experiments and complex systems.

Fault injection analysis based on simulation is widely used and accepted as an efficient way to perform soft errors assessment, enabling different system configurations explorations during early DSEs [26, 64]. Some state-of-the-art fault injectors use high-level behavioral models, and others are based on hardware description language level (HDL) or gate-level models. Although HDL-level and gate-level models are more precise than a high behavioral model, there are two main problems when using those models: First, commercial processors are rarely available to users in HDL or gate-level descriptions. Second, the simulation time in such abstraction levels is exceptionally high. Thus, thousands of simulations may take several months, which is not suitable to evaluate systems under soft errors during the early design phases. Given the ever-increasing complexity of both processor and software architectures, researchers have been investigating the use of virtual platforms as an alternative means to assess soft error resilience. The next chapter details the proposed use of virtual platforms as fault injection frameworks.

Chapter 3
Fault Injection Framework Using Virtual Platforms

This chapter describes the development of two fault injection frameworks designed on the basis of two virtual platform simulators: OVPsim (Sect. 3.5) and gem5 (Sect. 3.6). Additionally, Sect. 3.1 presents a brief description of the state-of-the-art on virtual platforms (VPs), while Sect. 3.2 displays a survey of VP-based fault injectors. After, Sect. 3.3 presents the adopted fault model. Section 3.4 describes the proposed fault injection flow integrated into both fault injection framework. Section 3.7 discusses the proposal of four novel fault injection techniques to improve the fault injection coverage. Performance improvements techniques are introduced in Sect. 3.8. Section 3.9 presents a set of modeling and simulation techniques developed to improve the performance and the design space exploration capabilities of both frameworks.

3.1 Virtual Platforms

A virtual platform is a full-system simulator that emulates hardware components (e.g., CPUs, memories), and the execution of real software stacks, on the same machine, as it is running on real physical hardware. Such simulators usually offer a set of processor architectures, peripherals, and memory models, allowing a fast and flexible software development at early design stages. VPs differ concerning modeling, flexibility, and simulation performance. While event-driven VPs such as gem5 target microarchitecture exploration, just-in-time (JIT) simulators (e.g., OVPsim [56]) are devoted to software development. JIT-based simulators can achieve speeds approaching actual execution time, e.g., thousands MIPS at the cost of limited accuracy. In turn, event-driven simulators typically report best-case simulation performances of 2–3 MIPS.

It is hard to cover all modeling aspects (e.g., flexibility, debuggability) and simulation (e.g., accuracy, scalability) requirements in one single simulator. This

F. Rocha da Rosa et al., *Soft Error Reliability Using Virtual Platforms*,
https://doi.org/10.1007/978-3-030-55704-1_3

Table 3.1 Most recognizable virtual platforms simulators

Simulator	ISA(s)	Guest OS	Accuracy	License
Simics	Alpha, ARM, MIPS, PowerPC, SPARC, ×86	Linux, Solaris, Windows, and others	Functional	Closed
PTLsim	×86	Linux	Cycle	GPL
Flexus	SPARC and ×86	Linux and Solaris	Cycle	BSD
SimpleScalar	Alpha, ARM, PowerPC, and ×86	Linux	Cycle	None
MARSS	×86	Linux	Cycle and Instruction	GPL
gem5	ARM, MIPS, SPARC, and ×86	Linux, Android, Solaris, and others	Cycle	GPL
QEMU	Alpha, ARC, ARM, PowerPC, MicroBraze, and others	Linux, Android, Solaris, and others	Instruction	GPL
OVP	Alpha, ARC, ARM, PowerPC, MicroBraze, and others	Linux, Android, Solaris, FreeRTOS, and others	Instruction	Modified Apache 2.0

section starts by providing an extension of the surveys [24, 48] considering the most popular virtual platforms. Such virtual platform simulators are compared according to different criteria: (1) accuracy, (2) flexibility regarding supported processor architectures, (3) licensing, and (4) support activity. Table 3.1 summarizes the reviewed work according to such four criteria.

The Wind River Simics [67] is a simulator that enables unmodified target software (e.g., operating system, applications) to run on top of a platform model. A wide range of processor architectures (e.g., ARM, MIPS, PowerPC), as well as operating systems (e.g., Linux, VxWorks, Solaris, FreeBSD, QNX, RTEMS), can be adapted to model the desired systems. This simulator includes SystemC interoperability, debuggers, software and hardware analysis views, as well. Simics is not open source, and thus users require a commercial license, which is one disadvantage.

The PTLsim [111] is a cycle-accurate simulator, offering the complete cache hierarchy, different processor architectures, memory subsystem, and hardware devices. This tool presents two main drawbacks: it only supports ×86-64 architectures, and it has no active maintained. SimpleScalar [7] is an open-source infrastructure for simulation and architectural modeling. Similar to previous simulators, software engineers can use SimpleScalar to develop applications and execute them onto a range of processor architectures, which varies from simple unpipelined processors to detailed microarchitectures with multiple-level memory hierarchies.

However, SimpleScalar is not actively maintained anymore (the last update was in March 2011), and other faster solutions, like gem5, are available.

QEMU is an instruction-accurate and open-source simulator that relies on dynamic binary translation supporting several CPUs (e.g., ×86, PowerPC, ARM, and SPARC). Similar to QEMU, the Open Virtual Platform Simulator (OVPsim) [56] engine also employs a just-in-time dynamic binary translation, i.e., target instructions (e.g., ARM, MIPS) are morphed into ×86-64 host instructions through a translation mechanism. The OVPsim API provides several component models, including processors, memories, UARTs, among others. The processor architectures and variants sum more than 170, including ARMv7, ARMv8, MIPS, Renesas, and MicroBlaze.

The gem5 [18] is a modular discrete event simulator, which has an open-source code and supports a rich set of models including processor cores, memories, caches, and interconnections. Among the available instruction set architectures (ISAs) are ×86, MIPS, Alpha, SPARC, and ARM (primarily architecture of this work) are examples of available instruction set architectures (ISAs). Further, the gem5 simulator is described in C++ and Python, and it has an active development and support team. It targets microarchitectural explorations, which incurs in substantial simulation overheads due to the number of modelled aspects (e.g., memories, caches, and interconnections). Further, the amount of memory required by the gem5 simulator is very high, making its use infeasible to explore large-scale system models. MARSS [81] is cycle-accurate full-system simulation of the ×86-64 architecture, which uses a hybrid approach through the PTLsim [111], as the basis of its CPU simulation environment on top of the QEMU [12].

QEMU was initially designed for single-processor platforms and virtualization purposes, and more recently, researchers are extending its capability to multicore explorations as well. However, the lack of documentation on the APIs or standardized methodology for creating manycore platform models limits its use. Excluding PTLSim and MARSS that only supports ×86, reviewed simulators provide support to several processor architectures. Cycle-accurate simulators such as SimpleScalar and gem5 entail high-simulation time, thereby limiting their applicability to more complex or large-scale systems explorations. Simics has a private license while the others are free to use. Further, SimpleScalar does not provide support or development anymore.

After careful selection, this work adopted OVPsim [56] and gem5 [18] as means to develop the fault injection frameworks. The OVPsim [56] is highly deployed in the industry and the academic communities under the frame of several research projects. The OVPsim can achieve speeds approaching thousands MIPS and its support several component models including 170 processor variants (e.g., ARMv7, ARMv8, MIPS, Renesas, MicroBlaze), memories, UARTs, among other components. Beyond the high-simulation performance, two main other reasons justify the adoption of OVPsim as means to develop the fault injection framework. First, among available VPs, OVPsim has the most substantial number of processor models and thus enabling a broader initial space exploration. Second, Imperas Software Ltd.[1] is directly involved in the work conducted in this work. One expected outcome of this collaboration is to promote the first commercial fault injection framework in the market. This work also adopts the state-of-the-art gem5 simulator [18] for three main reasons: (1) the gem5 source code is open and several extensions have been proposed in the past [3, 96], (2) the gem5 enables *microarchitectural cycle-accurate simulation* in an acceptable time (i.e., 0.4–2 MIPS depending on the application workload), and (3) it supports the current ARM Cortex-A architectures (i.e., ARMv7, ARMv8, and big.LITTLE).

3.2 Related Work on Fault Injection Approaches using Virtual Platforms

Virtual platform simulators facilitate fault injection implementation and analyses due to their design flexibility (e.g., several processor models available), and debugging capabilities (e.g., GDB support) are shown in Table 3.2. Authors in [49] present the Relyzer, a hybrid simulation framework for SPARC core using Simics [67] and GEMS [68] simulators coupled with a pruning technique to reduce injected faults. Also, the aforementioned framework fault injections target architectural integer registers and latches of the address generation units. In this work, a 200-cores cluster was employed and approximately 11 days were required to inject around 32 million faults, resulting in an average of 33 injected faults per second. The low number of injected faults is due to the high-cost simulation time of simics+GEMS simulator, which can achieve few hundred KIPS. This framework employs 12 benchmarks, four from each suite Parsec 2 [16], Splash-2 [108], and SPEC-Int [52]. In [92], Relyzer is extended to more aggressive pruning, reducing the number of faults that must be simulated. With the pruning and analysis techniques embedded in the GangES the authors further reduce this simulation time from 15,600 CPU-hours to 8200 CPU-hours.

In [28], a fault injection framework based on QEMU [12] is proposed. Faults are injected in an $\times 86$ architecture running applications in a Real-Time Operating System (RTEMS). The experimental setup accounts for four applications developed in-house. During the experiment, 8000 faults were injected in 8.7 h, given an average of less than one fault per second. The authors in [61] propose two tools: the GeFIN tool, a gem5-based fault injection framework and MaFIN, a MARSS-based [81] fault injection framework. In this work, faults were injected, randomly in time, in general-purpose registers, caches control registers, and other microarchitectural components. The experimental setup includes only the execution of 10 bare-metal benchmarks selected from the MiBench [47]. The use of an operational system is unknown.

---

[1] http://www.imperas.com/.

Table 3.2 Most recognizable virtual platform fault injection simulators

Info	Year			
	2014	2014	2015	2016
Authors	Hari et al. Relyzer	Geissler et al.	Kaliorakis et al. MaFIN and GeFIN	Guan et al. P-FSEFI
Simulator(s)	Simics GEMS	QEMU	MARSS gem5	QEMU
ISA(s)	SPARC V9	×86	×86 ARMv7	ARMv7 ×86
Applications	12: Parsec 2, Splash-2, and SPEC-Int 2006	Four in-house applications	10 MiBench	14 serial NASA NPB
Guest OS	OpenSolaris	RTEMS	None or unknown	None or unknown
Accuracy	Cycle and instruction	Instruction	Cycle	Instruction
Faults	32 M (15 K h)	32 K	300 K	140 K
Features	Architectural integer registers and in the output latches of the address generation units	Eight general-purpose registers, six segment registers, and instruction pointer	Architectural integer registers, L1 and L2 cache, and load–store queue	CPU logic units, registers, caches, and memory

Info	Year		
	2016	2016	2018
Authors	Didehban et al.	Tanikella et al.	This work
Simulator(s)	gem5	gem5	gem5 OVPsim
ISA(s)	ARMv8	×86 ARMv7 SPARC ALPHA	ARMV7 ARMV8
Applications	10 MiBench	10: MiBench and SPEC-Int 2006	26 OpenMP, 10 Serial, 9 MPI-Based and others
Guest OS	gem5 Syscall mode	None or unknown	Linux OSs, FreeRTOS, and Baremetal
Accuracy	Instruction	Cycle	Cycle instruction
Faults	72 K	33 K	3.3 M (1.2 M h)
Features	The register file, pipeline registers, functional units, and load–store queue	Eleven microarchitectural components	Register file and physical address space in both simulator. Virtual memory, variable data, function liveness, function instruction code in the OVPsim-based fault injection

Recently, in [101] the authors introduce the gemV, a gem5-based fault injection framework for microarchitectural elements such as instruction queue, reorder buffer, load–store queue, pipeline queue, renaming unit, and register file. The experimental setup includes ten application from the MiBench and SPEC-Int 2006 benchmark suites for eleven microarchitectural elements. Each element is subject to a 300-long fault campaign for each application, totaling 33,000.

The work [45] presents the P-FSEFI tool construct around the QEMU [12] simulator. This tool injects faults in the CPU logic units, registers, caches, and memory. The experimental setup consists of seven applications from the NAS Parallel Benchmark (NPB) [8], each one in a sequential and a parallel version; injecting 10,000 faults in each setup, totaling 140,000. The work in [27] introduces a gem5-based fault injection capable of flipping random register file bits, pipeline registers, functional units, and load–store queue. This work employs ten MiBench applications in a total of 72,000 faults.

Most reviewed approaches consider only small scenarios and only single-core processors. Exploration of soft error reliability of single-core architectures has been successfully supported over the last decades. However, the assessment of multicore architecture soft error resilience strongly requires complementary modeling and simulation mechanisms to manage other aspects such as resource sharing, memory allocation, and data dependencies. Further, such works typically report best-case simulation performances of 2–3 MIPS, allowing 33 fault injections per second considering a supercomputer [49].

3.3 Fault Modeling

This work reproduces the soft errors behavior using single-bit-upsets (SBUs) due to its higher probability of occurrence in electronic systems operating[58] at sea level. Our SBU model consists of a single bit-flip generated randomly in one general-purpose register (e.g., r0-13, sp, pc) or memory address during the application execution. Additionally, this model does not target the operational system (OS) explicitly (i.e., the boot process). Nevertheless, the OS calls arising in the application lifetime could be affected. This approach analyses the application behavior also considering the execution environment, thus exposing unforeseen consequences when compared with standalone implementation.

3.4 Fault Injection Flow

A group of fault injections targeting a particular application scenario including all related steps is here defined as a *fault campaign*. These steps are divided into four phases: the reference (or faultless) phase (1), fault generation (2), fault injection management (3), and final report generation phase (4). The fault campaign flow described by Fig. 3.1 is independent of the simulator. The simulation flow is implemented by the *simulation infrastructure (SI)*, which was mostly developed using shell and python scripts, thus being compatible with Linux, Windows, and MacOS OS distributions.

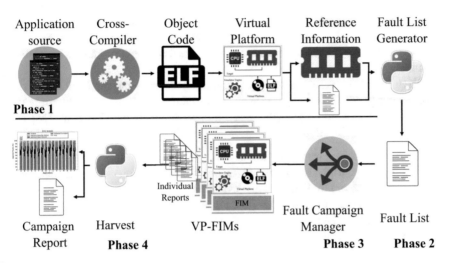

Phase 1

Individual Reports

FIM

Campaign Report   Harvest   VP-FIMs   Fault Campaign Manager   Fault List

Phase 4   **Phase 3**   **Phase 2**

Fig. 3.1 Fault injection campaign simulation flow

In phase one, the simulation infrastructure cross-compiles the application source for a given architecture resulting in the elf object file. After that, the SI simulates the application in an unmodified virtual platform (i.e., OVPsim or gem5) to verify its correctness and also to extract information using a gold standard and for fault creation. The register's contexts ((1), Fig. 3.1) and final memory state (2) compose this reference information. However, it can be easily adjusted to contain other types of fault injection (e.g., bus, network-on-chip, or cache information).

Phase 2 creates register or memory fault patterns consisting of injection time, a target, and a fault mask (i.e., the target bit). A random generation scheme selects the insertion time, the location, and the register bit since it covers the majority of faults at a low computational cost. The injection time ranges from one to the final instruction count extracted during phase 1 (Fig. 3.1). A bit-flip injection requires a bit mask with all bits set to "0," excepting the target bit. For instance, to change the second least significant bit of a given 32-bit register the following bit mask is required, 0×00000002. This fault generator can be extended effortlessly to include new types of fault or more sophisticated selection methods. Afterward, the complete fault list is available to the next phase from a simple plain text file.

Phase 3 (i.e., fault injections) is the most complex, and it's divided into two main sub-components: single fault injection and simulation speed boost. This flow description considers a single fault injection as performance boost techniques will be covered in Sect. 3.8. The simulation infrastructure maps one fault injection to one VP-FIM execution, in other words, ten fault injections require ten independent VP-FIMs. Each VP-FIM starts by reading the fault list and then schedule an event targeting insertion time (i.e., the number of executed instructions). At the fault injection moment, the FIM combines a fault bit mask with the current register value using an exclusive OR (XOR) operation. Supposing a 32-bit value of 0×00000009

and a fault targeting the fourth least significant bit, the module performs an XOR operation between 0×00000009 and 0×00000008, which generates 0×00000001. Subsequently, the FIM writes the new value in the target register and the simulation restarts.

Each VP-FIM instance performs an error analysis where the application behavior under fault injection is compared with the reference run (phase 1). The proposed flow supports custom error analysis and this work adopts the Cho et al. [26] five groups error classification:

- **Vanished**, no fault traces are left;
- **Application Output Not Affected** (ONA), the resulting memory is not modified. However, one or more bits of the architectural state is incorrect;
- **Application output mismatch** (OMM), the application terminates without any error indication. However, the resulting memory is affected;
- **Unexpected termination** (UT), the application terminates abnormally with an error indication;
- **Hang**, the application does not finish requiring a preemptive remove.

Lastly, phase four assembles all FIMs individual reports to create a single database and graphics.

3.5   OVPsim-FIM

The Open Virtual Platform simulator (OVPsim) [56] is an instruction-accurate simulator framework market by Imperas under an open license. The OVPsim relies on a dynamic binary translation (DBT) engine, which enables simulation running real application code at the speed of hundreds of MIPS. The DBT mechanism sequentially fetches each instruction from a target processor (or core), morphing it into new $\times 86\_64$ micro-operations. Further, it uses a micro-operations dictionary to hold the previous translations aiming to speed up the simulation process. The OVPsim multicore simulation algorithm creates fixed-length blocks of instructions belonging to each core. Each of those blocks is sequentially simulated. Note that in such simulation scenarios, each processor/core advances in fixed-length instruction steps. The OVPsim main limitation is the lack of detailed microarchitectural modeling (e.g., pipeline, decoder, reorder buffer) or cache interconnections. Reads and writes are always guaranteed to be atomic in an instruction-accurate simulator (i.e., the previous instruction always completes before the next starts), thus removing any data hazards originated from the pipeline access to information before the write-back stage update.

The fault injection module or just FIM is a series of components embedded in OVPsim, and it is responsible for:

- Monitoring the target processor;
- Accessing resources as memories and registers;

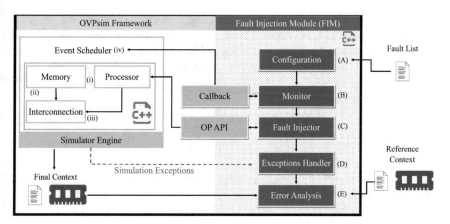

Fig. 3.2 OVPsim-FIM main components

- Injecting the faults;
- Capturing unexpected events arising from the simulator;
- Extracting information;
- Analyzing errors.

The developed FIM minimizes the intrusion in the simulator engine, thus enabling any researcher in possession of the original simulator to use, modify, or extend its functionalities. In a matter of fact, the OVPsim provides a complete set of APIs written in C/C++, which easily enables the simulator extension without any source code access. Figure 3.2 shows the OVPsim-FIM main components: (A) fault injector, (B) fault monitor, (C) configuration, (D) error analysis, and (E) exception handler.

The configuration component (C) reads the fault list, setups the monitor component (B), triggers the error analysis (D), and forwards the unexpected events to the exception handler (E), invoking each one at the appropriated time. We restrict this section to the essential features regarding the fault injection, and other available functions (e.g., speed boost techniques) are further discussed in the following sections. The monitor component (B) controls the internal simulator flow (i.e., starts and restarts the simulator), as the fault injection need to stop the simulator. This module schedules an event associated with the number of executed instructions to call the fault injector (A) whenever the injection time arrives. The fault injector component (A) access to the registers or memory through the OVPsim APIs enabling the modification of any available microarchitectural element. After the FI conclusion, it setups the hang (i.e., infinite loop) detector through a similar event targeting the stipulated threshold and resumes the simulation.

The exception handler (E) is responsible for capturing abnormal events arising from the application or OS (OS malfunctions, segmentation faults, or hard faults). Finally, the error detection component (D) verifies the simulator context for any mismatches considering the reference (faultless) execution. For this purpose, it extracts the current memory state, the register's context including the program counter, and the number of executed instructions. Thus, the error detection classifies the application under fault influence using the information acquired in the faultless execution. Also, it consolidates exceptions captured by the exception handler (E) to create a final report.

3.6   gem5-FIM

The gem5 simulator offers few accuracy levels depending on the exploration suitability, ranging from the simple atomic mode to a detailed one which includes an Out-of-Order (OoO) pipeline. The gem5 detailed mode provides a sophisticated memory timing and cache coherency protocols while the atomic mode emulates the memory and cache using a single cycle access mechanism. Therefore, the gem5 modes can slightly differ regarding the execution time and cache activity as additional cache misses or incorrect branch speculations can cause a pipeline flush. The gem5 simulator possesses an intermediary model with the atomic internal microarchitectural elements and enhanced memory access timing.

The unmodified gem5 simulator precision accuracy varies from 1.39 to 17.94% when considering the execution time against a prototyping board [24]. The microarchitectural elements do not always correlate to a particular hardware implementation due to speed overheads or to provide a more generic model. Thus, to address this issue, in [48] the gem5 was strictly modified to match the ARM Versatile Express board through the addition or modification of microarchitectural elements such as cache memory, branch predictor, and fetch buffer. This setup can achieve a mean percentage error smaller than 5% across several benchmarks.

The gem5 offers the complete source code and no proper extension API, and to maintain a non-intrusiveness extension, the gem5 fault injection module (gem5-FIM) independently extends the source code requiring only the access to some classes attributes/methods. The gem5 simulator employs Python scripts to control the simulation flow and C++ modules to model the microarchitectural components. Figure 3.3 displays the main gem5-FIM components, where white boxes represent the original gem5 components (i.e., (1) processor, (2) memory, (3) interconnection), and the blue the developed ones. The gem5-FIM incorporates five main components developed using Python: (A, Fig. 3.3) fault injector, (B) fault monitor, (C) configuration, (D) error analysis, and (E) exception handle and one C++ extension to access the microarchitectural components (F).

The gem5 has both a Python and C++ objects representations of components where the python configures and triggers the C++ objects through a series of events. Consequently, the gem5-FIM requires a C++ module alongside a Python counterpart. Conveniently the gem5 framework adopts the swig (i.e., Simplified Wrapper and Interface Generator) tool, which interfaces booth programming languages through wrappers. The Python side (A Fig. 3.3) gathers the processor and

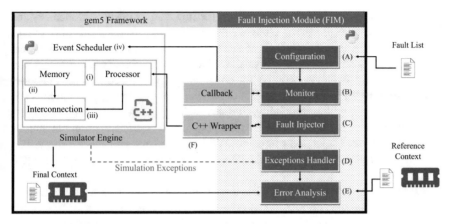

Fig. 3.3 gem5-FIM main components

memory handlers from the gem5 configuration and invokes the C++ wrapper (F)
to perform the bit-flip in the system description (e.g., memory, register file, pipeline)

3.7 Detailed Fault Injection Using Instruction-Accurate Virtual Platforms

Considering the works reviewed by Sect. 3.2, authors in [49] present the Relyzer,
a hybrid simulation framework for SPARC core using Simics [67] and GEMS [68]
simulators coupled with a pruning technique to reduce the number of injected faults.
The Relyzer is capable of injecting faults into architectural integer registers, and
output latches of the address generation unit. In [28], a QEMU-based fault injection
framework is proposed targeting general-purpose registers. Fault injection campaigns in this work consider an ×86 architecture running four in-house applications
on the top of RTEMS kernel. Another fault injection framework, called F-SEFI that
relies on QEMU is described in [45, 46]. The F-SEFI employs the QEMU using a
hypervisor mode, i.e., it does not emulate the complete target system, which reduces
both its fault injection and soft error analysis capabilities.

The authors in [61] propose the GeFIN and the MaFIN tools, which support the
injection of faults in microarchitectural components such as general-purpose and
cache control registers. Conducted experiments consider the execution of 10 bare-
metal benchmarks. Authors in [101] propose a gem5-based framework that allows
injecting faults in different microarchitecture elements (e.g., reorder buffer, load–
store queue, register file). In this work, each element is subject to small 300-long
fault campaign for each of the ten applications collected from both MiBench and
SPEC-Int 2006 benchmark suites. A similar gem5-based fault injection framework
is described in [27].

The reviewed frameworks only support the injection of bit-flips in memory and general single-core processor components, including registers, load–store queue, among others. Another drawback of such approaches is the lack of detailed and customizable post-simulation analysis. Reviewed works classify the detected soft errors according to the inspection of the processor architecture context (i.e., memory and registers), disregarding the impact of software components (e.g., functions and variables) on the system reliability. Further, such approaches typically report low simulation performances of up to 3 MIPS [49], which restricts the number and the complexity of fault injection campaigns. While some works consider a single ISA [49], others use only in-house applications [28] or bare-metal implementations [27, 61, 101].

Different from the above works, the OVPsim-FIM extension called SOFIA (*Soft error Fault Injection Analysis*) offers four novel non-intrusive fault injection techniques that provide software engineers with flexibility and full control over the fault injection process, allowing to disentangling the cause and effect relationship between a injected fault and the occurrence of possible soft errors, targeting an specific critical application, operating system or API structure/function. This tool also differs from all previous works by allowing users to define bespoke fault injection analysis and soft error vulnerability classifications, taking into account both software and hardware component particularities and the system requirements.

3.7.1   SOFIA

The SOFIA framework was developed on the basis of M*DEV simulator, a more advanced multicore and commercial version of the OVPsim. The M*DEV provides intercept libraries, and multicore debug primitives used to develop the SOFIA fault injection techniques. The SOFIA framework supports six fault injection techniques (A–F) making it suitable for fast and detailed soft error vulnerability analysis at an early design space explorations. Note that the *six* techniques target the register file or physical memory *without altering the target software stack* (i.e., application, OS, and related libraries). This extension proposes four new fault injection techniques besides the already available in the original OVPsim-FIM, Fig. 3.4 displays the six fault techniques supported by the module. These two techniques already embedded on the OVPsim-FIM randomly assign fault injections to deploy bit-flips targeting the Register File (A) (e.g., sixteen integer registers from r0 to r15) and Physical Memory (B) (e.g., one bit in a one-gigabyte memory), respectively.

3.7.1.1   Application Virtual Memory

The technique C targets the application virtual address space (VAS) to enhance the fault injection controllability. Operating systems abstract the physical hardware implementation from the user by making available a set of virtual address ranges

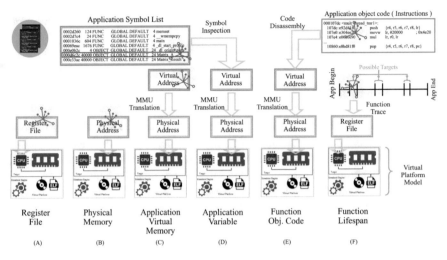

Fig. 3.4 OVPsim-FIM fault injection techniques according to with fault location and injection time

while using a translation table to connect both virtual and physical ranges. The promoted technique C automatically extracts the virtual addressing ranges from the target application object code, including different segment addresses (e.g., data, code, read-only, debug) during the phase 1 in order to create the fault lists (phase 2). For each fault injection, the SOFIA accesses the target OS virtual memory translation table, acquires the correspondent physical address from the target virtual address, and injects the bit-flip in the system physical memory. The advantage of this technique over the purely physical memory fault injection relies on the fact that it targets the application virtual address space (VAS) without affecting the OS, other applications, or libraries, reducing the number of necessary faults campaigns since soft errors manifest much quicker. Additionally, this approach enables the user to target a particular application running in a complex environment with multiple applications and libraries.

3.7.1.2 Application Variables and Data Structures

To precisely evaluate an application vulnerability to soft errors the fault injection infrastructure should provide efficient means to correlate errors with particular application blocks or data structures. Technique D (Application Variable) enables the software engineer to direct bit-flip injections into particular data structures, enabling to isolate and identify the most vulnerable ones with a lower number of fault campaigns and higher precision. Further, this approach allows evaluating the impact of specific application variables on the soft error reliability without affecting the application main control flow. For this purpose, the user is asked to inform the target variable name, and the SOFIA framework automatically captures the variable virtual address to create a set of faults targeting the data structure virtual addressing. During the application execution, the variable will suffer a single bit-flip on its physical memory representation using the aforementioned translation table.

3.7.1.3 Function Code

To explore the criticality of function codes, this work proposes the technique E (*Function Object Code*) that limits the injection spectrum to the memory region, which holds the target function code (i.e., instructions), including local variables, etc. The probability of function code to be hit by a transient fault depends on its relative size when compared to the complete memory range. This technique enables the user to investigate the soft error reliability of a particular function independent of its size or execution time.

3.7.1.4 Function Lifespan

In state-of-the-art frameworks, the fault injection time follows a random generation scheme where faults are scattered over the entire application and OS execution. Consequently, the number of faults per function depends on its execution time and not on its criticality for the system reliability. *Function Lifespan* (F) technique enable to reduce the fault injection spectrum by limiting the insertion time to those small intervals where the target function is active. During the simulation, the fault monitor component (B, Sect. 3.2) traces the function execution at the instruction level and thus create a list of active ranges, including the processor core(s) that executed the underlying function. The lifespan fault injection technique targets the general-purpose registers (r0-r15), the program counter (PC), and the stack pointer (SP). However, this technique can be combined along with any other fault injection technique (e.g. C, D, and E) whenever necessary.

3.7.1.5 Fault Inspection

The OVPsim-FIM extension provides a flexible soft error assessment module, which enables the creation of customizable error classifications. The software engineer can alter the classification order, add new classes, change their criteria, or include new parameters. This module is capable of extracting one or more variables value on-the-fly during the simulation, and compare then with pre-characterized data. For instance, a target variable can be compared to another variable in the same application. New classifications are easily embedded in the fault campaign flow, enabling custom fault injection scenarios according to the application requirements. Injecting a bit-flip in the physical memory (i.e., B, C, D, and E) produces dirty memory in the majority of the cases. In other words, the targeted bit remains untouched until the simulation conclusion, resulting in silent data corruption (SDC).

Nevertheless, if the mismatch against the faultless execution is the target bit (i.e., no additional control flow or memory errors), it can be considered benign depending on the system constraints. The SOFIA tool is capable of differentiate dirty memories from other types of errors.

3.8 Improving Fault Injection Campaigns Performance

A fault injection campaign comprises thousands of simulations, demanding a significant computational effort, limiting most investigations to small or simpler scenarios. This work investigates the virtual platforms as a technique to speed up early design space explorations regarding reliability assessment of commercial processors. Although some VPs simulation speed reaches hundreds of MIPS, simulating complex software stacks is still a real challenge. Two characteristics are noticeable by observing the fault injection campaign: (1) One fault injection is independent of the other. (2) The application executes for a significant time without the influence of faults (i.e., equal to gold execution) to provide an appropriated context for the fault injection. This code executed before the fault injection is unnecessary and do not influence the final fault result. The proposed fault campaign deploys techniques to reduce the simulation computation cost at different levels of granularity, which are discussed in the next subsections.

3.8.1 Shared Memory Multicore Parallelization

The fault injection campaign is naturally a parallel process as the fault injections are independent of each other. To exploit this characteristic, the simulation infrastructure deploys multiple VP-FIMs across a shared memory multicore processor, nowadays found in any workstation. The simulation infrastructure automatically limits the maximum number of platforms running in parallel to match the number of host cores as shown in Fig. 3.5. Each platform matches one Linux process with an overall core utilization of 100%, thus in a quad-core, the simulation infrastructure allows four platforms in parallel. The simulation infrastructure surveys the platform process identification (PID) number and dispatches the next platform shortly after the completion of any running VP-FIM.

3.8.2 Checkpoint and Restore Technique

The checkpoint technique consists of periodically saving the system context during a faultless execution and later restoring the appropriate context for each fault injection. During faultless execution, the VP-FIM stores periodically (i.e., according

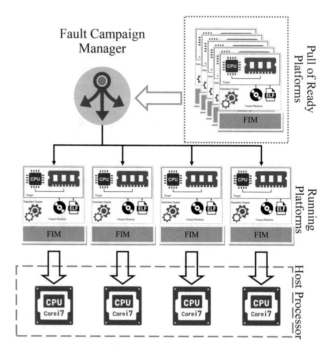

Fig. 3.5 Host multicore fault injection campaign

to a predefined instruction interval) the application context covering processor and memory models. At the fault campaign, each FIM identifies the closest checkpoint before the fault injection time to be restored. Additionally, the fault injection event trigger adjusts the injection time considering the fast for the number of forwarded instructions. The user can specify this interval or assign some checkpoints, and thus, the simulation infrastructure automatically estimates the interval between checkpoints.

The OVPsim-FIM creates the checkpoints concomitantly with the faultless execution. For this purpose, the OVPsim-FIM includes a checkpoint component responsible for stopping and saving the context according to the specified interval by the user. During the fault campaign, the FIM selects the best matching checkpoint to restore (i.e., the first checkpoint before the fault injection time). From this point, the injection process resembles the version without checkpoints. While the OVPsim provides memory and processor checkpoints, it does no offers the same functionalities for other modules (e.g., UARTS, timers). Consequently, the deployed checkpoint mechanism current works only in bare-metal platforms.

The gem5 simulator supplies save and restore functionalities, which allows restoring processor and memory context from a binary file. A restored processor model should execute identically to an unmodified execution, nevertheless, the gem5 does not always behave as expected. The gem5 checkpoint function has a significant limitation, as it automatically restarts the simulation engine to process

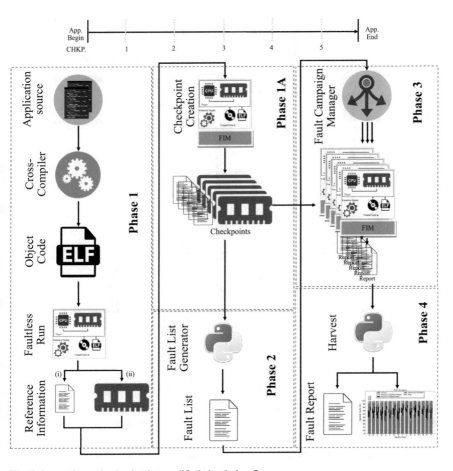

Fig. 3.6 gem5 atomic checkpoint modified simulation flow

pending events. This action introduces few hundreds of ticks[2] and in some occasions, increases the simulation in a few instructions. Resuming, a simulation run without using checkpoints executes slightly fewer instructions (and ticks) when compared with a simulation recovered from checkpoints. This behavior does not impact the application behavior, however, to avoid any mistaken comparison it is necessary to extract the exact information. The simulation flow was modified to incorporate a checkpoint profiling run (see Phase 1A Fig. 3.6) to overcome this problem. This phase extracts the checkpoints and at the end updates the reference information (i.e., faultless). At this point, the simulation flow has two reference data sets: One without checkpoints and another after the checkpoint process. Consequently, the FIM can compare the correct set during the error analysis.

---

[2]One tick is the minimum granularity inside the gem5, and usually, each clock cycle has 500 ticks.

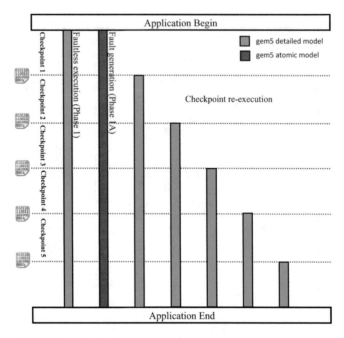

Fig. 3.7   gem5 detailed mode checkpoint scheme

The gem5 detailed mode has another peculiarity (see Fig. 3.7), the simulator until the present moment does not create executable checkpoints (i.e., which cannot be restored). The new checkpoint scheme first creates the checkpoints using the gem5 atomic mode, as previously described (Phase 1A). Later each checkpoint will be simulated using the detailed mode until the application end to acquire the reference information. Assuming one application with five checkpoints, for instance, first the faultless execution (Phase 1), then a checkpoint generation using the atomic (one simulation). Finally, the gem5-FIM detailed simulates the five checkpoints At this point, we have six possible references: One without a checkpoint and five for each checkpoint re-execution. Additionally, this phase occurs in parallel to reduce the time overhead.

3.8.3   Distribute Fault Injection Campaigns Using HPCs

High-performance computers had been used for many decades by researchers of many domains such as quantum mechanics, weather forecast, environmental research, and oil and gas explorations to speed up simulations. HPC system can be divided into two distinct classes: Extremely complex applications which require a considerable amount of resources (e.g., RAM, storage, cores) not available in standard workstations. The other type consists of more straightforward and smaller applications easily executable in ordinary computers, however, which requires a significant number of single executions to obtain meaningful results.

OVPsim is suitable for few thousand faults when considering the contemporary workloads with hundreds of billions of instructions. For example, a single execution of the largest application (i.e., NPB EP) requires 12 h or 4000 days in an 8000-fault scenario using the gem5 atomic. This work proposes a simulation infrastructure extension to support distributed fault injections across an HPC system (i.e., University of Leicester ALICE supercomputer). The ALICE has 170 standard nodes; each one has two 14-core Intel Xeon Skylake, 128 GB of RAM, and a local storage disk. Further, this HPC extension uses the Portable Batch System (PBS) framework, which is used in most larger scale systems to provide a generic solution.

An HPC environment deploys a job scheduler to maximize the overall hardware utilization where a job (i.e., shell scripts that executes a specific work) describes the necessary resources (i.e., memory, the number of cores, the number of nodes, and wall time precisely) and commands. The walltime (i.e., the maximum execution time) is the job most significant attribute, and usually, the job scheduler deploys a First-come, First-served service policy with some modifications. HPC systems aim for higher hardware utilization and not always the fairest resource division, consequently smaller jobs jump ahead to fill gaps in the scheduling. Longer jobs will wait in the queue longer, and to reduce the starvation problem the scheduler also deploys a priority inversion policy.

The simulation flow follows the previous sections and Fig. 3.8 illustrates the same flow across multiple computer nodes. The first phase is performed only once (A, in Fig. 3.8), where the application and disk image are compiled in a local computer and then transferred to the supercomputer due to some environmental limitations. The

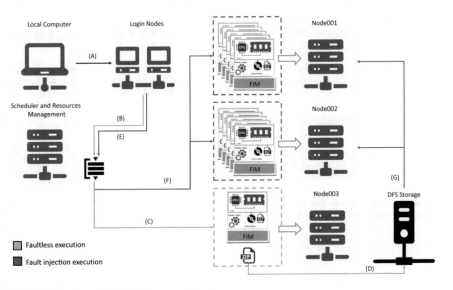

Fig. 3.8 Distributed fault injection campaign flow

reference phase (B, Fig. 3.8) collects the reference information and generates the fault list. Additionally, this job estimates the total simulation time for the scenario (i.e., the number of faults times the execution time of one fault injection) and selects the number of jobs to be submitted.

The simulation infrastructure does not match one fault per job as it would create a substantial management overhead, and instead, it agglomerates the fault blocks in approximately 24-h jobs.[3] Considering an 8000-fault campaign and an application simulation time of 1 h, the optimum arrangement in this scenario requires 24 applications per job (i.e., a walltime of 24 h) requiring 333.33 jobs, which is unfeasible. The simulation infrastructure selects the best fit between the number of faults per job and walltime, where for the previous example is 400 jobs with 20 h walltime. Therefore, the simulation infrastructure creates job templates (i.e., bash scripts) for each fault injection scenario.

Developing an application to HPC systems present several challenges, for example, the number of jobs running in parallel is limited by the distributed file system performance. The simulation flow has three options to manage scenarios source files (e.g., simulator executable, checkpoint files, application binaries, fault list):

1. The easiest (i.e., no significant modification is needed) solution executes remotely over the distributed file system (DFS) where the HPC remotely access files in the storage using the network. This option only fits small scenarios (up to 25 jobs) due to the bottleneck created in the DFS by the multiple access to the same memory region.
2. The next option transfers all simulation-related files across the network to the node local storage during each job startup. Nevertheless, the initial copies may overlap with other jobs reducing the overall transference speed and congesting the DFS buffers due to the number of files required (i.e., thousands of files). Using the local storage improves the flow scalability to approximately 500 parallel jobs, not satisfying our requirements.
3. The last solution compresses the simulation source files in a single zip file during the phase 1, copying this file during each job startup to the node storage where they are locally decompressed. This approach enables the management of thousands of concurrent jobs by simulation infrastructure, and it is the adopted in this work.

The jobs responsible for the fault injection are submitted to the scheduler and queued (E, Fig. 3.8) to later be assigned to nodes as the resources are made available (F). During the job initialization, it copies the compressed file (G) (simulation source files), locally extracts the contents, and the node executes the designated platforms. The individual reports are later merged into a final fault injection report.

---

[3]Jobs with walltime up to 24 h are classified as short jobs by the ALICE scheduler and thus increasing the execution possibility.

3.9 Other Extensions

State-of-the-art software stacks, including OSs, compilers, and application work-loads leading to several extensions were conducted in both FIM simulators.

3.9.1 Targeting Complex SW Stacks

The proposed FIM and surrounding simulation infrastructure described a general methodology to inject faults in the register file of a given processor until this point. Beyond controlling the fault campaign flow as described in Sect. 3.4, the simulation infrastructure requires other modifications to include an operating system. First, any OS needs a bootloader,[4] kernel image, file system image, and a different cross-compiler. The simulation infrastructure automatically appends the application binary file in the file system image, selects the bootloader and kernel files. Indeed, the development time spends on the development and improvement of the simulation infrastructure is comparatively more substantial than the FIM by itself.

The VP-FIM should be able to capture the exact moment where the application finish (i.e., *main* function return statement). Also, the FIM captures application starting point (i.e., faults are only injected during the application execution) and abnormal events (i.e., attempt to execute an illegal instruction or out of range memory address). In bare-metal systems this can be easily accomplished, for example, surveying physical memory access. However, OS abstracts and protects the hardware components from the guest applications requiring a new solution.

These issues lead to the development of an OS/VP-FIM communication API called *FIM-API*, enabling the application inside the guest OS to call different VP-FIM services. Among the available services are: the application begin and end point, a segmentation fault events, and others. The gem5-FIM version makes use and extends the gem5 built-in artificial instructions to deploy these necessary services. Note that the artificial instructions are implemented using a library linked to the application, and thus, requiring no modification in the compiler. The OVPsim provides an ISA-independent and more generic solution as it is capable of creating virtual memory callbacks on-the-fly. The OVPsim-FIM intercepts special function symbols by name during the system execution. The simulation infrastructure links an FIM-API header file together with the application while the Linux file system image includes an FIM-API standalone executable.

---

[4]Some OSs require more than one bootloader.

3.9.2   Injecting Faults on Multicore Systems

The adopted VP-FIM handles an arbitrary number of cores/processors limited only by the simulator supported architectures. To inject and collect the results from multiple cores, few modifications are necessary: During the reference phase (1), the FIM collects individual core information such as instruction count and register state. The fault creation takes into account the target architecture core count, as it distributes the faults evenly through available cores. For instance, when creating 8000 fault campaign targeting a quad-core ARM processor, the fault creator will assign 2000 faults for each core. Also, it holds specific information concerning each core instruction count necessary for the hang threshold. After the simulation end, it retrieves the information to compare with the faultless execution for each core. The complete simulation infrastructure automatically adjusts the FIM, fault generator, and other components according to the user core selection.

3.9.3   ARMv8 Architecture Extension

The development of more complex applications in the mobile domain and the user demand for high-performance devices lead to the adoption of 64-bit architectures [62, 107]. This new architecture enables a larger virtual memory addressing alongside other microarchitectural enhancements. The ARM company developed a series of processors (e.g., Cortex-A53, Cortex-A72) fitted with a new AARCH64 instruction set. This new ISA had 33 general-purpose registers and fixed 32-bit instruction width, restricted conditional execution instructions to branches, alongside new float-pointing, and encryption support. The AARCH64 reduces and simplifies the previous ARMv7-A ISA, removing old legacy instruction and executions modes particularly suitable for embedded controllers.

The VP-FIM extension to support the new ARM architecture requires modifications in the FIM and simulation infrastructure. This ISA has 17 new integer register, totaling 33 64-bit wide registers. The reference phase and fault injection capture this new registers for error analysis. Also, the fault generator has 64-bits extensions to target the entire architecture. The simulation infrastructure requires additional Linux kernels, file system images, makefiles, and support libraries.

3.10   Closing Remarks

This chapter presented a novel fault injection framework which targets commercial multicore system executing complex software stacks. The promoted framework provides a fast and flexible soft error assessment tool, especially for early design space explorations. During this phase, multiple software stacks and hardware configurations can be tested in feasible time using the two VP-FIMs developed in this chapter. After exploring randomly assigned fault injections this work investigated new techniques to target/expose the system critical software segments. In this way, it enables a more substantial fault coverage, whenever targeting a complex software stack, by prioritizing the critical portions of the code. Next, Chap. 4 will use this FI framework to evaluate hundreds of distinct fault injection scenarios considering commercial multicore systems and real workloads.

Chapter 4
Performance and Accuracy Assessment of Fault Injection Frameworks Based on VPs

This chapter investigates the performance and the soft error evaluation accuracy of two fault injection frameworks developed on the basis of two well-known VPs: OVPsim and gem5. Underlying investigation considers an extensive comparison between two distinct virtual platforms (i.e., instruction- vs. cycle-accurate) regarding simulation performance, soft error analysis credibility/accuracy, and fault injection flexibility; For accomplishing these three objectives, this chapter comprises 3,344,000 fault injections which require up to 2 million simulation hours. Considering a single-thread sequential computer, this workload requires approximately more than 150 years. Table 4.1 summarizes the fault campaign presented in this chapter. Next, Sect. 4.1 describes the adopted benchmarks characteristics.

4.1 Experimental Setup

This work performs thousands of fault injections using distinct application. Configurations including single-, dual-, quad-, and octa-core ARM Cortex-A9 processors (ARMv7 Architecture) or Cortex-A72 (ARMv8). The gem5-FIM includes a two-level cache memory model where the detailed mode provides a more accurate timing model, while the OVPsim does not account for a cache memory model, and, thus, accessing the RAM directly. To avoid external influences and assure the closest comparison, the software stack uses the same compilation environment regarding compiler, flags, libraries, and target an identical Linux kernel. Note that the OVPsim does not currently possess a checkpoint in place for the Linux platform, and instead, it boots the Linux kernel for each fault injection. However, the checkpoint load process would be less efficient than simulating the entire boot due to the high OVPsim simulation speed. Consequently, the OVPsim simulation time accounts for the kernel startup, which increases the simulation to approximately 1.3 billion of instructions. Also, the OVPsim-FIM observes when the application

F. Rocha da Rosa et al., *Soft Error Reliability Using Virtual Platforms*,
https://doi.org/10.1007/978-3-030-55704-1_4

Table 4.1 Fault injection campaigns summary for the explorations presented in Chap. 4

VP	Description	Scenarios	Total of faults	Simulation time (h)
OV	16 Rodinia benchmarks for one, two, and four cores	48	384,000	545.82
GA	16 Rodinia benchmarks for one, two, and four cores	48	384,000	5855.72
GD	16 Rodinia benchmarks for one, two, and four cores	48	384,000	44117.78
OV	10 NPB OpenMP version for one, two, and four cores	30	240,000	1599.22
OV	10 NPB Serial version for a single core	10	80,000	490.69
OV	9 NPB MPI version for one, two, and four cores	25	200,000	1160.11
GA	10 NPB OpenMP version for one, two, and four cores	30	240,000	179137.78
GA	10 NPB Serial version for a single core	10	80,000	394468.89
GA	9 NPB MPI version for one, two, and four cores	25	200,000	578555.56
OV	16 Rodinia benchmarks for one, two, and four cores; time-slice 0.00001	48	384,000	615.53
OV	16 Rodinia benchmarks for one, two, and four cores; time-slice 0.000001	48	384,000	782.86
OV	16 Rodinia benchmarks for one, two, and four cores; time-slice 0.0000001	48	384,000	815.00
		Total	3,344,000	1208144.96

begins, which enables it to inject fault only after the OS complete boot. Table 4.2 summarizes the VP-FIMs experimental setup which is identical to perform a fair comparison between multiple scenarios. Further, MPI applications require a local communication library, which should be compiled and included in the Linux virtual disk. This work uses the MPICH, a high-performance and widely portable implementation of the Message Passing Interface (MPI) standard [44].

This work adopts two distinct workloads:[1] the Rodinia benchmarks [25] and the NASA NAS Parallel Benchmarks (NPB) [8]. The Rodinia is a set of well-known applications developed by the University of Virginia aiming the high-performance computing domain. This suite assembles 24 parallel applications using three different programming APIs (i.e., OpenMP, CUDA, and OpenCL) from distinct domains such as Medical, Biological, Physical, Data Mining, and Image Processing. Nevertheless, CUDA and OpenCL are GPU-based programming languages, and so requiring a distinct simulator, which does not belong to the scope of this work. For this experimental setup, we select 16 OpenMP benchmarks as shown in Table 4.3

---

[1] This work adopts benchmark and application as similar terms.

Table 4.2 Virtual platforms experimental setup

Parameter		OVPsim	gem5 atomic	gem5 detailed
Architecture	ARMv7	Cortex-A9 multicore		
	ARMv8	Cortex-A72 multicore		
Memory	RAM	One gigabyte of RAM		
	Cache	None	L1 Inst 32kB 4-way associative L1 data 32 kB 4-way associative L2 512 kB 8-way associative	
Cross-compiler	ARMv7	arm-linux-gnueabi-gcc Ubuntu 6.2.0-5ubuntu12		
	ARMv8	aarch64-linux-gnu-gcc Ubuntu 6.2.0-5ubuntu12		
Compilation flags	ARMv7	-O3 -g -w -gdwarf-2 -mcpu=cortex-a9 -mlittle-endian -DUNIX -static -fopenmp -pthread		
	ARMv8	-O3 -g -w -gdwarf-2 -mcpu=cortex-a72 -mlittle-endian -DUNIX -static -fopenmp -pthread		
Linking flags	ARMv7	-static -fopenmp -lm -lstdC++ -lm5		
	ARMv8	-static -fopenmp -lm -lstdC++ -lm5		
OS	ARMv7	Linux Kernel 3.13.0-rc2		
	ARMv8	Linux Kernel 4.3.0+		

Table 4.3 Selected Rodinia applications

#	Name	Domain	#	Name	Domain
A	Backprop	Pattern recognition	I	Myocyte	Biological simulation
B	bfs (breadth-first search)	Graph algorithms	J	nn (k-nearest neighbors)	Data mining
C	Heartwall	Medical imaging	K	nw (Needleman–Wunsch)	Bioinformatics
D	Hotspot	Physics simulation	L	Particle filter	Medical imaging
E	Hotspot3d	Physics simulation	M	Pathfinder	Grid traversal
F	kmeans	Data mining	N	srad v1	Image processing
G	lavaMD	Molecular dynamics	O	srad v2	Image processing
H	lud	Linear algebra	P	Streamcluster	Data mining

from A to P. Table 4.4 shows the simulation time in seconds (i.e., Intel Core I7-7700K 4.20 GHz with 16 GB DDR4 2400 MHz) for the Rodinia applications varying the number of target cores and VP-FIM: OVPsim-FIM, gem5-FIM atomic, and gem5-FIM detailed.

The Rodinia simulation time using the OVPsim-FIM is around six seconds independently the number of target cores due to the just-in-time engine. In contrast,

Table 4.4 Rodinia
applications simulation time
varying the VP and number
of cores

	Simulation time (s)								
	OVP			gem5 atomic			gem5 detailed		
	Cores			Cores			Cores		
#	1	2	4	1	2	4	1	2	4
A	4	5	4	11	11	11	30	33	37
B	6	6	6	24	26	27	124	136	150
C	7	5	5	17	17	17	93	103	116
D	5	5	5	22	22	25	111	120	137
E	6	6	6	73	73	75	536	500	526
F	4	4	4	14	14	15	56	60	74
G	4	6	5	20	19	21	123	123	147
H	5	6	5	13	12	14	36	40	60
I	5	4	4	14	15	17	66	71	97
J	6	6	8	57	70	104	359	645	1410
K	4	5	4	13	14	15	48	59	79
L	4	4	4	39	42	42	282	313	506
M	4	4	4	14	15	15	73	74	76
N	4	5	5	76	71	73	467	507	566
O	4	4	4	40	40	42	260	284	313
P	5	5	5	53	66	59	445	516	746

the gem5 has several interconnect models for different hardware components, and, therefore, each application behavior (e.g., the number of instructions, type of instructions, memory accesses pattern, cache misses, and branch predictor misses) impacts simulation time. Some application (e.g., J, L) are visibly impacted by the number of target cores, for example, the application J simulation time using the gem5 detailed mode varies from 359 to 1410 s, in this case, due to its memory access pattern (data mining algorithm).

The NAS Parallel Benchmark is developed by the NASA Advanced Supercomputing Division [8] as a set of programs designed to evaluate the performance of parallel supercomputers. With constant support, these applications suffered several revisions during the last years to correct errors and improve its performance. Further, this suite has a unique feature among the benchmark suites: some parallel applications (i.e., OpenMP and MPI) derive from a common serial version. Table 4.5 shows the application name, description, and parallelization paradigm. The NPB aims modern high-performance computers leading to a larger workload when compared with the Rodinia. For example, the Rodinia applications simulation time ranges from 4 to 104 s (i.e., considering the gem5 atomic only), while the NPB varies between 176 and 42,370 s (i.e., 12 h) for a single simulation. Due to the more extended applications, the NAS exploration uses only the gem5-FIM atomic mode because the detailed mode has unacceptable simulation times (i.e., more than a day per simulation). We chose to execute the serial version using a single-core ARM Cortex-A9, as its lacks any explicit parallelization. Table 4.6 displays the simulation

Table 4.5 NAS parallel benchmarks

Name	Description	Serial	OpenMP	MPI
BT	Block tri-diagonal solver	×	×	×
CG	Conjugate gradient, irregular memory access and communication	×	×	×
DC	Data cube	×	×	
DT	Data traffic			×
EP	Embarrassingly parallel	×	×	×
FT	Discrete 3D fast Fourier transform, all-to-all communication	×	×	×
IS	Integer sort, random memory access	×	×	×
LU	Lower-upper Gauss–Seidel solver	×	×	×
MG	Multi-grid on a sequence of meshes, long- and short-distance communication, memory intensive	×	×	×
SP	Scalar penta-diagonal solver	×	×	×
UA	Unstructured adaptive mesh, dynamic and irregular memory access	×	×	

time for Serial, MPI, and OpenMP application variants considering one, two, and four cores systems.

4.2   Performance and Accuracy Evaluation of Instruction-Accurate Virtual Platforms

This section compares the OVPsim-FIM (instruction-accurate) precision against the gem5-FIM (cycle-accurate) considering three main aspects: accuracy (Sect. 4.2.1), OVPsim engine configuration (Sect. 4.2.2), and simulation speed (Sect. 4.2.3). For this purpose, this study employs the ARM Cortex-A9 ISA configured in the OVPsim-FIM and gem5-FIM using atomic and detailed modes. Recapping, the deployed soft error model consists of randomly generated single bit-flips injected in any available general-purpose register (i.e., r0–15) during the software stack execution (i.e., OS, drivers, and applications). The OS reliability is not the primary focus of this chapter, and thus fault injections only occur during the application lifespan (i.e., the OS startup is not subject to faults). Nevertheless, OS system calls arising during this period (i.e., application execution time) are susceptible to fault injections as part of the application behavior.

Table 4.6 NPB benchmarks simulation time in seconds

	Simulation time (s)					
	OVP			gem5 atomic		
	Cores			Cores		
App	1	2	4	1	2	4
Serial applications						
BT	36.7	a	a	5.12×10^3	a	a
CG	19.7	a	a	1.92×10^3	a	a
DC	13.2	a	a	182	a	a
EP	157	a	a	3.95×10^3	a	a
FT	31.9	a	a	5.22×10^3	a	a
IS	13.3	a	a	163	a	a
LU	22.9	a	a	2.21×10^3	a	a
MG	13.9	a	a	312	a	a
SP	21.7	a	a	2.67×10^3	a	a
UA	114	a	a	23.3×10^3	a	a
OpenMP-based applications						
BT	47.5	39.6	43.8	5.10×10^3	5.48×10^3	6.28×10^3
CG	20.1	21.8	24.5	1.85×10^3	2.06×10^3	2.52×10^3
DC	25.1	15.0	16.5	176	234	225
EP	185	181	179	3.81×10^3	42.4×10^3	42.2×10^3
FT	33.1	34.5	43.3	5.55×10^3	5.93×10^3	6.05×10^3
IS	13.6	14.7	16.1	171	186	202
LU	23.6	24.0	26.5	2.28×10^3	2.49×10^3	3.26×10^3
MG	14.3	15.5	18.3	318	340	357
SP	22.1	24.0	34.4	2.75×10^3	2.83×10^3	3.48×10^3
UA	124	126	138	24.9×10^3	26.5×10^3	27.2×10^3
MPI-based applications						
BT	38.1	a	51.5	5.00×10^3	a	6.26×10^3
CG	21.3	22.9	29.2	18.4×10^3	2.06×10^3	2.49×10^3
DT	23.9	15.1	18.5	440	417	466
EP	151	155	168	39.0×10^3	41.3×10^3	42.4×10^3
FT	33.9	36.2	47.4	5.48×10^3	5.85×10^3	6.39×10^3
IS	14.0	21.3	29.6	176	225	472
LU	26.0	26.1	29.7	2.66×10^3	2.99×10^3	3.59×10^3
MG	14.7	16.8	19.4	332	391	619
SP	23.4	a	29.5	2.90×10^3	a	3.87×10^3

[a]Not available scenarios

4.2.1   Accuracy

This subsection explores the impact on the soft error vulnerability assessment using different software simulation approaches: discrete event-driven simulation (e.g., gem5) and a just-in-time dynamic binary translation engines (e.g., OVPsim). Figures 4.1 and 4.2 show 8000 randomly assigned register fault injections for each scenario using 10 NPB Serial, 10 NPB OpenMP, 9 NPB MPI, and 16 Rodinia OpenMP applications.

The gem5 has a higher vanished percentage (i.e., no trace emerges from the fault injection), in particular, the detailed mode when compared against the OVPsim. This difference can be traced to components only available in the detailed mode leading to microarchitectural masking, i.e., some hardware component overwrites the target register/memory bit before the fault propagation to other memory elements. For instance, the gem5 detailed mode has a more precise cache coherency model, increasing the cache miss rate slightly, and by consequence, re-fetching some cache lines. The register renaming module is component exclusive to the detailed mode that protects the register file from pipeline data hazards by mapping logical registers (e.g., r0–15) on physical ones. Both OVPsim and gem5 (atomic mode) do not emulate these two microarchitectural elements, then using single-cycle read and write operations resulting in the similar masking rate. In contrast, applications G, I, and N (Fig. 4.1a) are exceptions of this behavior, in these cases, the OVPsim and gem5 (detailed mode) have more similar outcomes than the gem5 atomic versus the detailed. Due to its instruction-accurate engine the OVPsim simulation time (i.e., number of executed instructions) is affected by the running application characteristics (Sect. 4.2.2 details this behavior). In turn, the vanish errors results collected from gem5 detailed show that a more substantial occurrence of ONA error when compared to gem5 atomic and OVPsim-FIM, as illustrated in Fig. 4.1c by applications such as C and M executing on a quad-core processor.

The Rodinia benchmarks have a more significant presence of ONA errors than the NPB, in other words, at least one incorrect register bit (e.g., r1, PC, SP) differs from the faultless execution. The NPB longer execution reduces the probability of dirty bits (i.e., lower ONA presence) in the simulation outcome due to a higher likelihood of a bit masking when comparing with the Rodinia applications. For example, the NPB applications BT and EP (Fig. 4.2a) or the Rodinia benchmarks A, F, and L (Fig. 4.1b) when using the OVPsim. Further, the Rodinia OpenMP applications have a higher number of hangs than the NPB ones, notably increasing the number of cores (Fig. 4.1c). A *Hang* occurs when the application execution time exceeds the double of expected time (i.e., time compared to the faultless executions). The two leading causes to explain the higher hang presence in the Rodina: (1) the fault affected a loop statement (e.g., while, for) wherein these cases a more extended execution translates to more significant recovery time. For comparison's sake, the average Rodinia application executes 80 million instructions, while NPB applications execute on average around 17 billion of instructions (i.e., $212\times$ larger). (2) kernel malfunctions: the fault injection leads to unrecoverable kernel perturbations (e.g., a thread scheduler error). A longer execution time reduces the Linux kernel exposure time (i.e., the probability of kernel function be stroke by a fault). In other words, the more prolonged the applications, the less kernel functions execute proportionally.

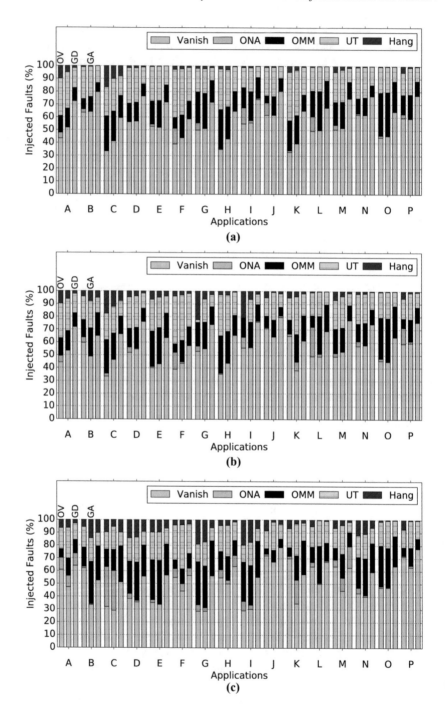

Fig. 4.1 Rodinia benchmarks 8000-fault injection campaign for a multicore ARM Cortex-A9 processor. (**a**) Single-core ARM Cortex-A9 processor. (**b**) Dual-core ARM Cortex-A9 processor. (**c**) Quad-core ARM Cortex-A9 processor

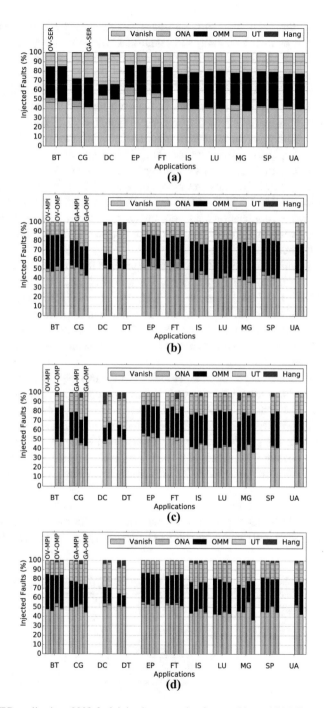

Fig. 4.2 NPB applications 8000-fault injection campaign for a multicore ARM Cortex-A9 processor. (**a**) Serial applications. (**b**) MPI and OpenMP applications using the single-core processor. (**c**) MPI and OpenMP applications using the dual-core processor. (**d**) MPI and OpenMP applications using the quad-core processor

When considering a multicore processor, increasing the core count results on more thread context switching and combined with sub-linear scalability (i.e., underutilized cores in this context) from the Rodinia applications leads to further the kernel errors. During CPU inactivity moments the OS executes the scheduler algorithm[2] and then moves to a sleep mode *wait for interruption*. Faults striking during this waiting period will remain in the core register file until its wake-up, affecting the Linux kernel thread dispatcher and system control flow.

The gem5 suffers event scheduler malfunctions in some specific cases due to unforeseen application behavior resulting in a simulator crash (classified as UTs). The gem5 simplistic memory representation as a single binary vector, and whenever the Linux MMU translates an out of range address, the simulator reaches a segmentation fault in the host system. On the other hand, the OVPsim withstands better to unexpected application behaviors and continues to simulate the application. Thus, it exceeds the predefined threshold to be considered in an infinite loop, and the FIM pronounces it as a hang error.

To facilitate the data comprehension, we introduce the *Accumulated Classification Mismatch* (ACM), which is defined as the sum of absolute differences between classes divided by the total number of faults. The two VP-FIM in a hypothetical three classes case study (X, Y, and Z) under comparison as presented on Table 4.7. The difference between classes is (i.e., 5, 5, and 10), 20 from 150 fault injections. Thus, the accumulated classification mismatch for this scenarios ten divide by 150 equals to 6.66%.

In order to analyze the differences among the VP-FIMs Figs. 4.3 and 4.4 explore the accumulated classification mismatch in three comparisons: gem5-FIM atomic versus gem5-FIM detailed, OVPsim-FIM versus gem5-FIM atomic, and OVPsim-FIM versus gem5-FIM detailed. As aforementioned, the NPB experimental setup does not include the gem5-FIM detailed due to its higher simulation time, and, thus, the NPB mismatch considers only the OVPsim-FIM versus gem5-FIM atomic. Tables 4.8 and 4.9 summarize the ACM information in average, worst case, and best case.

Figure 4.3a compares the gem5 atomic and detailed modes using the Rodinia applications in a multicore system (i.e., one, two, and four cores). Excluding the application C (heartwall) where the mismatch increases from 25 to 38%, the number

Table 4.7 Accumulated classification mismatch hypothetical 150-faults scenario

Class	VP-FIM 1	VP-FIM 2	Absolute difference
X	40	35	5
Y	60	55	5
Z	50	60	10
Total absolute difference			20
Total accumulated mismatch			10
Total relative difference			6.66%

---

[2]This scenario executes one application per time, and, thus, there is no other thread to run.

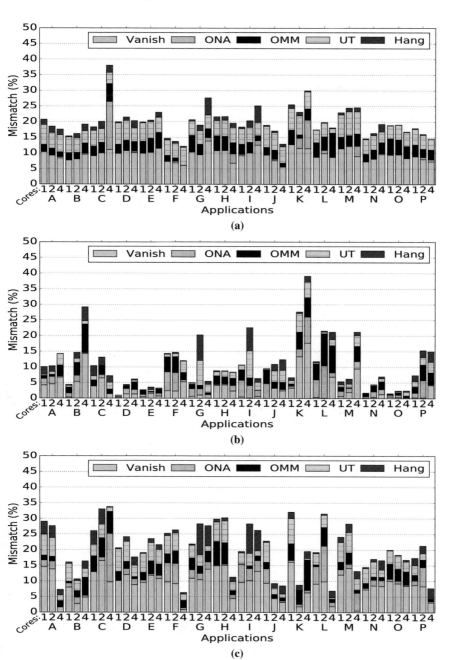

Fig. 4.3 Rodinia benchmarks ACM considering a multicore ARM Cortex-A9. (**a**) gem5-FIM atomic vs. gem5-FIM detailed. (**b**) OVPsim-FIM vs. gem5-FIM atomic. (**c**) OVPsim-FIM vs. gem5-FIM detailed

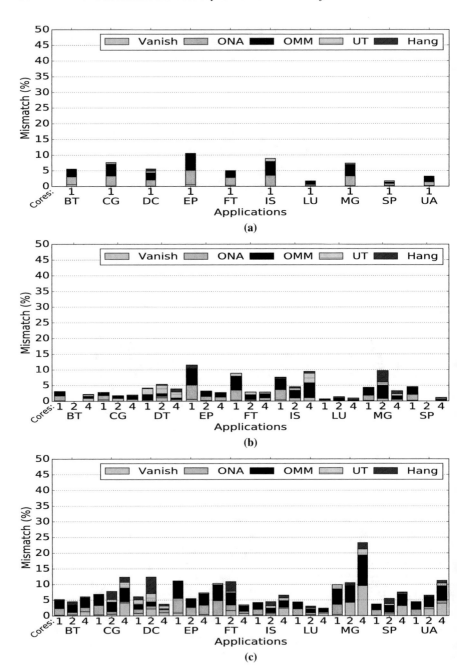

Fig. 4.4 NPB applications ACM between the OVPsim-FIM and gem5-FIM atomic considering a multicore ARM Cortex-A9. (**a**) Serial applications. (**b**) MPI applications. (**c**) OpenMP applications

Table 4.8 Rodinia benchmarks ACM summary comparing three distinct VP-FIMs

#	Comparison	One core (%)	Two cores (%)	Four cores (%)
Worst case	OV vs. GA	14.34	27.64	39.07
	OV vs. GD	32.12	33.02	33.79
	GA vs. GD	25.50	24.45	38.01
Best case	OV vs. GA	0.94	2.43	2.46
	OV vs. GD	14.49	8.70	6.28
	GA vs. GD	14.65	13.71	12.15
Average	OV vs. GA	6.95	12.55	13.17
	OV vs. GD	22.32	22.91	16.08
	GA vs. GD	19.11	19.20	21.17

Table 4.9 NPB ACM summary considering the OVPsim-FIM against the gem5-FIM atomic

#	API	One core (%)	Two cores (%)	Four cores (%)
Worst case	Serial	10.55	a	a
	MPI	11.55	9.72	9.49
	OpenMP	11.15	12.36	23.25
Best case	Serial	1.70	a	a
	MPI	0.68	1.36	0.96
	OpenMP	3.69	3.05	2.38
Average	Serial	5.73	a	a
	MPI	5.32	3.21	3.17
	OpenMP	6.62	7.11	8.39

[a]Not available scenarios

of cores has little impact. For example, the application A (backprop) mismatch reduces with a growing number of cores, while the benchmark I (myocyte) has the opposite behavior. In other words, the atomic mode lack of some microarchitectural components imposes an almost constant difference when compared with the detailed mode.

The OVPsim-FIM mismatch against the gem5-FIM detailed is around 22% (i.e., single- and dual-core), while 6.95 and 12.55% when compared with the gem5-FIM atomic (Table 4.8). Expected behavior when comparing the instruction-accurate OVPsim against a microarchitectural simulator due to the already discussed microarchitectural masking mechanisms. However, the quad-core scenario shows a lower average mismatch from the OVPsim vs. detailed than atomic vs. detailed. In this case, both OVPsim and gem5 detailed execute more instructions then the gem5 atomic, however, for distinct reasons. The gem5 detailed mode has a better memory timing model adding cycles to the execution time, and, thus, delaying the OpenMP synchronization events. The OVPsim engine uses a block-based approach to serialize the multicore simulation adding waiting times between inter-core synchronizations used by the OpenMP.

Figures 4.3b and 4.4a–c show several comparison scenarios between the OVPsim-FIM and gem5-FIM atomic. OpenMP applications show a mismatch worsening while increasing the number of cores, for example, the Rodinia experiments B, D, and K (Fig. 4.3b) alongside the NPB application MG and CG (Fig. 4.4c). The Rodinia mismatch increases using quad-core comparing with single-core processors with the worst case jumping from 14.34 to 39.07%. In the same context, the average error grows from 6.95 to 12.55% considering one and two cores and remains stable for four cores with a mismatch of 13.17%. The ACM has a more diverse distribution across the categories than the gem5 atomic versus gem5 detailed, for example, in some applications, the vanish classification has a more significant mismatch and in others the ONA.

Also, NPB longer workloads reduce the ACM in general when compared with the Rodinia suite. NPB OpenMP applications average mismatch varies from 6.12 to 8.51% in contrast with the Rodinia figures of 6.95 and 13.17%. The worst-case mismatch between the gem5 atomic and OVPsim reduces up to 60% when using the NPB applications compared with the Rodinia benchmarks. MPI applications have in general a smaller mismatch for all experiments, and for instance, the worst case reduces from 23.25 to 9.49% considering a quad-core processor. Serial, MPI, and OpenMP differences will be further explored in Sect. 5.1.2. Next Sect. 4.2.2 explores different OVPsim engine configurations under fault injection.

4.2.2   Instruction-Accurate Simulation Engine Parameters Impact on Soft Error Assessment

The previous subsection explored the soft error analysis accuracy of the OVPsim-FIM instruction-accurate framework against cycle-accurate gem5. Results show an average error of 11% in all cases (considering the gem5 atomic), with the worst case achieving up to 40%. Especially, the OpenMP applications displayed a more significant mismatch than the MPI or serial ones. This subsection investigates the origin of such mismatch and analyzes some solutions to improve the OVPsim-FIM accuracy. First, it is necessary to understand how the gem5 and OVPsim software simulation approaches behave under fault presence. The gem5 describes the target microarchitecture as components (i.e., register file, pipeline, cache) interconnected by a series of events. A scheduler in the gem5 engine executes these events at each simulation tick, updating the whole system state including multiple cores, memories, and other subsystems. Events are executed at a rate of 500 ticks per CPU cycle in the simulated system, and consequently, a complete instruction takes a couple of thousands of ticks.

The OVPsim relies on just-in-time (JIT) dynamic instruction translation engine, which translates the target ISA (e.g., ARMv7, ARMv8) to host $\times$86-64 instructions, providing its higher simulation speed. Further, a complete instruction is the OVPsim minimal simulation granularity, in other words, the simulation always advances one instruction. Similarly to an OS scheduler where several processes share the same CPU time, the OVPsim engine simulates each model instance (i.e., processor, core, peripheral) for a fixed-length block of instructions called *Quantum*. The quantum size is configurable using a variable, *time-slice*, representing a time in seconds.[3] The quantum size is given by the following equation where by default the time-slice is 0.001 s (1 ms):

$$Block\ Size = (processor\ nominal\ MIPS\ rate) \times 1E6 \times (time\text{-}slice\ duration)$$
$$(4.1)$$

The target processor nominal MIPS rate is 448 MIPS resulting on a quantum size equal to $448 \times 1E6 \times 0.01 = 448,000$ instructions. The OVPsim also deploys a scheduling policy to manage the processors and other components simulation. The simulator selects the first processor, after the first processor (or core) has simulated during 448,000 instructions, it is suspended, and the next processor assumes. In the case of multicore processors such as the ARM Cortex-A9×2, each processor core receives a separate quantum, and it is scheduling accordingly. Figure 4.5 displays two simulation scenarios regarding a dual-core processor, one with the default quantum and another using half of its size (i.e., 224,000 instructions), which increases the number of model switches for an identical workload. This block simulation approach delays inter-core communication (or synchronization events), as the sender and receiver cores cannot execute simultaneously. For instance, communications between the first and the fourth cores (i.e., considering a quad-core processor) take at least two full quantums because the OVPsim needs to simulate second and third quantum blocks, before simulating the target fourth core.

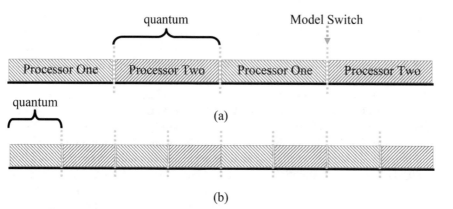

Fig. 4.5 OVPsim scheduling policy varying the quantum size for a dual-core processor executing the same workload. (**a**) Default quantum size—448,000 instructions. (**b**) Smaller quantum size—224,000 instructions

---

[3]The time-slice value in seconds refers to an internal configuration parameter and not to the simulation, host, or real-time.

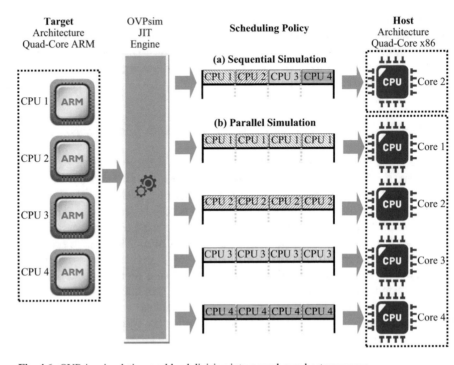

Fig. 4.6 OVPsim simulation workload division into a quad-core host processor

To diminish the soft error mismatch between the gem5 and the OVPsim it is necessary to reduce the time required to complete the inter-core communications. Considering the JIT-engine characteristics, there are two main solutions: reduce the time-slice variable or parallelize the quantum simulation. In this context, the OVPsim provides an acceleration feature called *quantum-leap* (QL), which enables mapping processor models (e.g., Quad-core ARM) to physical host cores (i.e., ×86_64 multicore). For example, considering a quad-core ARM processor model, in the sequential simulation mode (Fig. 4.6a), four target cores (CPU 1–4) share a single ×86 host core (Core 2). In turn, in the quantum-leap mode (Fig. 4.6b) each ARM model (CPU 1–4) is individually simulated in an ×86 host core (Cores 1–4).

First, this work explores the impact of using distinct quantum sizes (i.e., 448,000, 4480, 448, and 44 instructions per block) and quantum-leap configurations on the OVPsim-FIM accuracy to assess soft error reliability, considering gem5-FIM as the reference. Further, the two quantum-leap scenarios (i.e., QL-1 and QL-2), where each one employs a distinct thread allocation scheme restricted to quad-core processors. Table 4.10 shows the six proposed scenarios mismatch between OVPsim-FIM and gem5-FIM.

Table 4.10 Rodinia benchmarks quantum explorations worst, best, and average cases considering the gem5-FIM atomic as reference

#	Time-slice	One core (%)	Two cores (%)	Four cores (%)
Worst case	0.001	14.33	27.63	39.07
	0.00001	12.32	19.86	13.40
	0.000001	12.02	19.68	11.97
	0.0000001	11.31	18.53	11.42
	QL-1	a	a	71.24
	QL-2	a	a	38.69
Best case	0.001	0.93	2.42	2.46
	0.00001	1.12	1.91	1.30
	0.000001	0.52	1.10	1.72
	0.0000001	0.60	1.55	1.23
	QL-1	a	a	4.00
	QL-2	a	a	3.84
Average	0.001	6.95	12.55	13.16
	0.00001	5.79	9.17	5.54
	0.000001	4.96	8.95	5.23
	0.0000001	4.09	7.62	5.39
	QL-1	a	a	17.56
	QL-2	a	a	13.38

[a] Not available scenarios

4.2.2.1   Quantum-Leap Impact on Soft Error Assessment

The first scenario (QL-1, Table 4.10) uses a greedy allocation algorithm and affinity thread, i.e., children threads can only execute in a fixed physical core to avoid host caches synchronizing costs. In this context, the OVPsim QL children threads compete with other system threads depending on the system current workload. In the second scenario (QL-2, Table 4.10) the thread affinity is relaxed, enabling children threads to execute in any host core and apply a less aggressive thread allocation technique, which reduces the resource competition with other system threads.

QL-1 and QL-2 scenarios show a worse accuracy for all applications when compared to the single-core OVPsim-FIM execution, highlighting that the only distinction between both scenarios is the thread allocation scheme. The number of quantums executing in parallel varies according to the target application and host system workload leading to a simulation performance improvement of 200% on average when using a host quad-core processor. The quantum-leap provides a higher performance, however, its disadvantages outweigh the simulation performance gain considering the soft error analysis. First, soft error analysis requires deterministic simulation of thousands of FI campaigns and the QL execution can be affected by the host OS thread allocation mechanism and current workload. Second, the FI flow enables other better speedup techniques as discussed in the previous chapter. From this point, this work will focus on the sequential execution of the OVPsim only (Fig. 4.7).

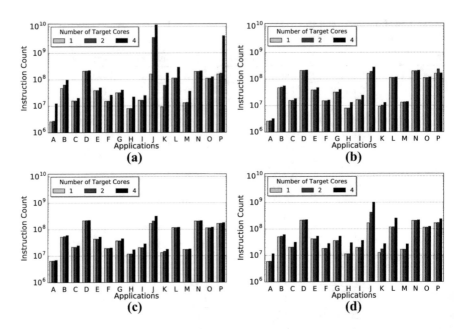

Fig. 4.7 Instruction count for a single faultless application execution considering four VP-FIMs using one, two, and four cores. (**a**) OVPsim default quantum (448 K instructions). (**b**) OVPsim 44-instruction quantum. (**c**) gem5 atomic. (**d**) gem5 detailed

Table 4.11 Rodinia mismatch benchmarks comparison considering the OVPsim-FIM with default (DF) and 44-instruction quantum (Q) against the gem5-FIM atomic

#	Comparison	One core (%)		Two cores (%)		Four cores (%)	
		DF	Q	DF	Q	DF	Q
Worst case	OV vs. GA	14.34	11.31	27.64	18.54	39.07	11.43
	OV vs. GD	32.12	33.55	33.02	30.74	33.79	39.16
	GA vs. GD	25.50		24.45		38.01	
Best case	OV vs. GA	0.94	0.60	2.43	1.55	2.46	1.24
	OV vs. GD	14.49	14.41	8.70	16.57	6.28	8.54
	GA vs. GD	14.65		13.71		12.15	
Average	OV vs. GA	6.95	4.10	12.55	7.62	13.17	5.40
	OV vs. GD	22.32	21.54	22.91	24.18	16.08	22.38
	GA vs. GD	19.11		19.20		21.17	

Red means larger mismatch than the DF
Green means smaller mismatch than the DF

4.2.2.2   Mismatch Considering the Quantum Size

Besides the Table 4.11, Figs. 4.8a, 4.9a, and 4.10a show the reference gem5-FIM atomic (ψ) and four quantum sizes 448,000 (λ), 4480 (γ), 448 (β), and 44 (δ) instructions per block for one, two, and four cores on an ARM cortex-A9, respectively. The experiments show that the quantum reduction has a significant impact on

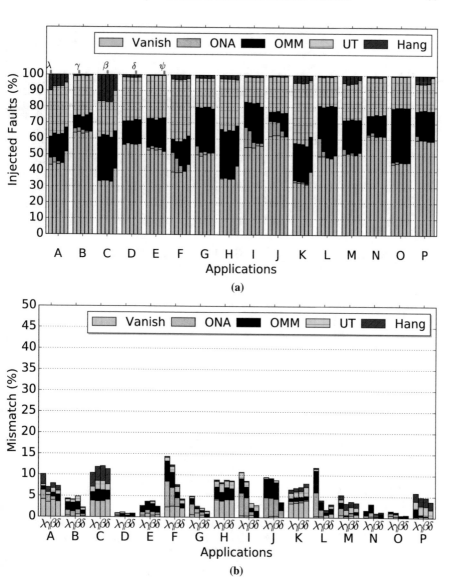

Fig. 4.8 8000-fault injection campaign deploying the OVPsim-FIM for a single-core ARM Cortex-A9 processor varying the time-slice and the reference gem5-FIM atomic. gem5-FIM atomic (ψ) and OVPsim-FIM time-slices values: 0.001 (λ), 0.00001 (γ), 0.000001 (β), and 0.0000001 (δ). (**a**) Fault campaign. (**b**) Accumulated classification mismatch for each OVPsim-FIM time-slice value

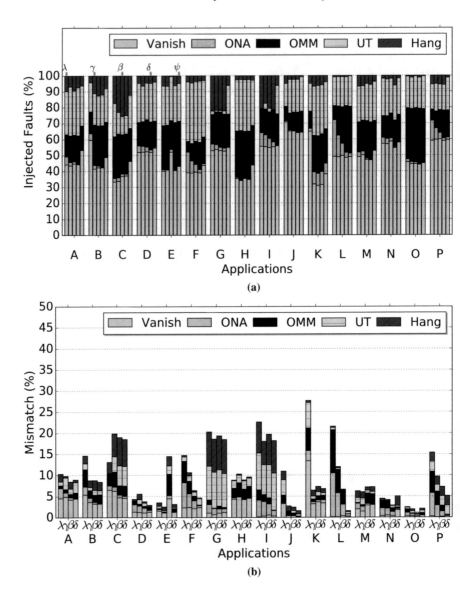

Fig. 4.9 8000-fault injection campaign deploying the OVPsim-FIM for a dual-core ARM Cortex-A9 processor varying the time-slice and the reference gem5-FIM atomic. gem5-FIM atomic (ψ) and OVPsim-FIM time-slices values: 0.001 (λ), 0.00001 (γ), 0.000001 (β), and 0.0000001 (δ). (**a**) Fault campaign. (**b**) Accumulated classification mismatch for each OVPsim-FIM time-slice value

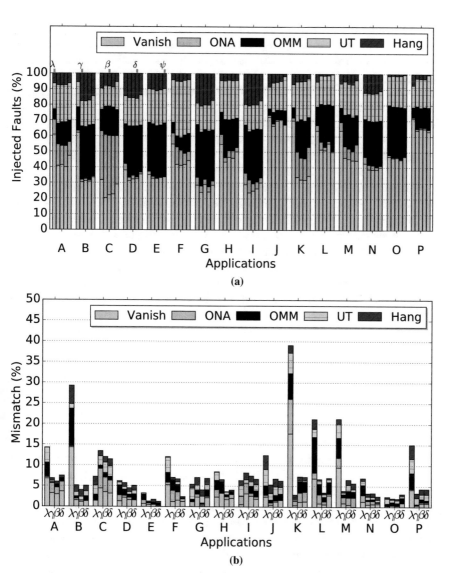

Fig. 4.10 8000-fault injection campaign deploying the OVPsim-FIM for a quad-core ARM Cortex-A9 processor varying the time-slice and the reference gem5-FIM atomic. gem5-FIM atomic (ψ) and OVPsim-FIM time-slices values: 0.001 (λ), 0.00001 (γ), 0.000001 (β), and 0.0000001 (δ). (a) Fault campaign. (b) Accumulated classification mismatch for different OVPsim-FIM time-slices for a quad-core ARM Cortex-A9 processor in comparison with the gem5 atomic

the soft error analysis of OVPsim-FIM. For example, the quad-core processor model presents an average accuracy improvement of up to 40% when using the smallest block (i.e., 44 instructions), while some benchmarks reach a fivefold accuracy gain (e.g., F and I). The quantum 10,000 smaller (i.e., 44 instructions) cuts the average error from 6.95 to 4.09% in the quad-core scenario (Table 4.10) while the *worst case* reduces from 39.07 to 11.42%. Noticeably, reducing the quantum size decreases the communication cycles between cores approximating the OVPsim-FIM and gem5-FIM behaviors. While the smallest block size presents the best accuracy, using a slightly larger quantum of 4480 instructions leads to 86% of worst-case improvement and 91% in the average mismatch.

The resulting mismatch can be traced back to its block-based simulation engine, as previously discussed, each core executes a fixed amount of instructions before changing to the next one. Note that inter-core communications are completed during the core switch, leading to temporally unsynchronized cores. Inter-core communication is necessary to synchronize events across multiple cores, for instance, in a parallelization library. The Rodinia OpenMP-based applications use a fork–join parallelization paradigm where synchronization barriers coordinate multiple children threads execution. One synchronization event that requires all cores to reach the same statement (i.e., a barrier) requires multiple quantum executions until completion. Delaying these communication events lead to some extra instructions executed by the OVPsim due to other cores waiting.

It is possible to observe this behavior by comparing the instruction count of different VP-FIMs executions. Figure 4.7 displays the number of executed instructions for a single faultless execution considering the OVPsim with the default quantum (a), a 10,000 smaller quantum (b), the gem5 atomic (c), and the gem5 detailed (d). Note that in the quad-core scenario the applications *nn, nw, streamcluster, and backprop* present a more substantial variance in the number of executed instructions at the same time as they present the worst soft error mismatch. In contrast, when the overall OVPsim using the 44-instructions quantum follows closely the gem5 atomic comportment (see Fig. 4.7c). For instance, in the *nw* scenario, the OVPsim-FIM executes nine times more instructions than gem5-FIM atomic with a 78.15% soft error mismatch. The same scenario using the smaller quantum results on only 28% more executed instructions along with a fivefold soft error accuracy improvement. Reducing the quantum size diminishes this inter-core communication gap (i.e., where one core waits for another) and approaches the gem5-FIM atomic moded behavior under FI, especially when targeting multicore architectures. While the optimal quantum size varies according to the application behavior and how its synchronization primitives are defined, its reduction improves the overall simulation accuracy.

We select the smallest quantum (i.e., a 44 instructions block size) to expand our investigation using the NPB suite. Figure 4.11 shows the Rodinia benchmarks FI using the OVPsim-FIM with reduced quantum (OV-Q), the gem5-FIM atomic (GA), the and gem5-FIM detailed (GD). Figure 4.12 replicates the previous experimental setup using the NPB. Tables 4.10 and 4.12 display the mismatch between gem5-FIM and OVPsim-FIM using the default quantum size (DF) of 448,000 instructions and a

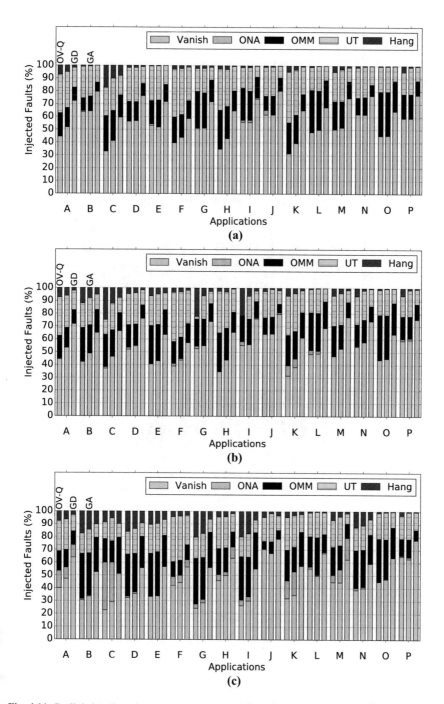

Fig. 4.11 Rodinia benchmarks using the OVPsim-FIM with reduced quantum (OV-Q), the gem5-FIM atomic (GA), the and gem5-FIM detailed (GD) for a multicore ARM Cortex-A9 processor. (**a**) Single-core ARM Cortex-A9 processor. (**b**) Dual-core ARM Cortex-A9 processor. (**c**) Quad-core ARM Cortex-A9 processor

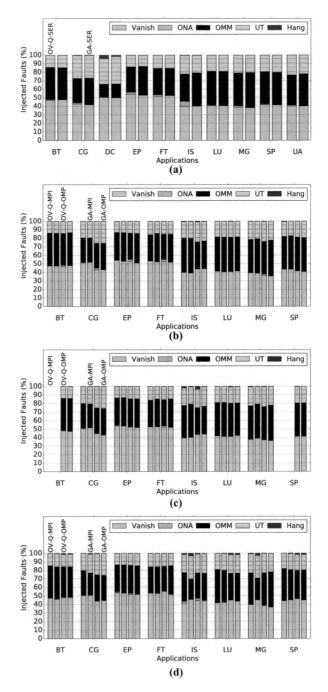

Fig. 4.12 NPB applications using the OVPsim with smaller quantum (OV-Q) and gem5-FIM atomic (GA) for a multicore ARM Cortex-A9 processor. (**a**) Serial applications. (**b**) MPI and OpenMP applications using the single-core processor. (**c**) MPI and OpenMP applications using the dual-core processor. (**d**) MPI and OpenMP applications using the quad-core processor

Table 4.12 NPB mismatch comparison considering the OVPsim-FIM with default (DF) and 44-instruction quantum (Q) against the gem5-FIM atomic

#	API	One core (%)		Two cores (%)		Four cores (%)	
		DF	Q	DF	Q	DF	Q
Worst case	Serial	8.90	7.34	a	a	a	a
	MPI	11.55	3.45	9.72	3.29	9.49	12.50
	OpenMP	10.32	4.38	12.36	4.21	23.25	4.50
Best case	Serial	1.70	0.79	a	a	a	a
	MPI	0.68	0.16	1.36	0.55	0.96	1.34
	OpenMP	3.69	1.04	3.05	0.84	2.38	0.24
Average	Serial	5.20	2.73	a	a	a	a
	MPI	5.32	1.30	3.21	1.36	3.17	4.18
	OpenMP	6.12	2.61	7.29	2.07	8.51	2.10

[a]Not available scenarios

smaller quantum (Q) value of 44 instructions, considering four distinct workloads: NPB OpenMP (OMP), NPB MPI, and NPB Serial (SER) alongside the Rodinia OpenMP applications.

Notice, the OVPsim-Q (i.e., OVPsim with the 44-instructions quantum) decreases the mismatch whenever compared with the gem5-FIM atomic for the Rodinia suite, and more accentuated in the quad-core system due to the instruction count reduction as previously mentioned. For example, the average ACM reduces by 58% (from 13.17 to 5.4%) as shown in Table 4.10. In the same scenario, the worst-case scenario diminishes from 39.07 to only 11.43%. Note the migration of ONA to OMM by decreasing the quantum size, in other words, the incorrect content previously restricted to the register file migrates to the final memory. For instance, note application A, E, F, and L in Fig. 4.11a. Notice, the campaigns simulated with the OVPsim-Q presents a mismatch reduction in 28 out of 30 scenarios with a significant (five-times) improvement in the worst case of OpenMP-based benchmarks, which is justified by the impact of synchronization barriers between children threads.

While the Rodinia benchmarks include applications with up to 220 *million*, NPB benchmark applications vary from 16 to 87 *billion* instructions. By consequence, NPB benchmarks have more extended computations between synchronization points than the Rodinia, which impacts on the soft error analysis. NPB benchmarks also have a better workload distribution and scalability, which means in conjunction with the more prolonged execution that children threads have enough instructions to complete one or more quantums between OpenMP barriers. In contrast, Rodinia applications have a shorter computation time between synchronization points, sometimes smaller than one complete quantum execution. This behavior leads the OpenMP barriers to execute additional instructions while waiting for other threads (i.e., delays for an extra quantum at least). The Rodinia behavior magnifies the mismatch originated due to the OVPsim simulation policy using fixed-length instructions blocks, and by consequence, these applications benefit the most when reducing the quantum, achieving a fivefold accuracy gain. Applications using the MPI library are based on independent threads, which leads to a smaller number of synchronization points, and thus resulting in a lower mismatch. The quad-core MPI workload has two NPB applications (*IS and MG*) in which the scenarios with RQ lead to a mismatch worsening. As discussed before, some application may suffer from over reduced quantum size, for instance, by selecting the best case for these two applications; the average mismatch reduces to 3.48 and worst case of 5.85%. *IS and MG* are the smallest NPB applications that might contribute to the precision worsening by reducing the number of instruction per quantum.

The OVPsim-Q decreases the mismatch considering the gem5 atomic while increasing when compared with gem5 detailed. For example, the OVPsim-Q quad-core scenario average ACM is 22.38% is higher than previously 16.08% with the unmodified OVPsim engine (Table 4.10). Even though the unmodified OVPsim-FIM more accurately emulates the gem5-FIM detailed in some cases, this setup increases the mismatch. The OVPsim-Q has a more similar relationship when compared with the gem5 atomic, and by consequence, a better average figure can be extracted.

4.2.3   Performance and Speedup

The OVPsim has two main advantages when comparing with other frameworks: modeling flexibility and simulation speed. This subsection compares the simulation speed regarding MIPS considering the gem5 and OVPsim in multiple configurations and workloads. The experiments were conducted in a Quad-core Intel Core I7-7700K 4.20 GHz with 16 GB DDR4 2400 MHz. Figure 4.13 presents the simulation performance as we increase the host cores from 1 to 4 where both the OVPsim-FIM and gem5-FIM can perform and manage parallel simulations.

Figure 4.13a displays the first set of simulation considering the Rodinia applications. The gem5-FIM atomic simulation speed ranges from 4.2 to 11 MIPS, while the detailed mode achieves 0.89–1.65 MIPS. In turn, the OVPsim varies from 345 to 2921 MIPS depending on quantum configuration and application. Modifying the quantum size reduces the number of executed instructions per block impacting simulator performance directly due to the increasing switching between the models (i.e., cores). The unmodified OVPsim-FIM (in gray Fig. 4.13a) has an average performance of 1500 MIPS, the 4400 (red) and 440 (blue) reduce the average performance in 30 and 42%, respectively, while the smallest quantum reaches 360 MIPS. The second experiment, Fig. 4.13b, deploys the NPB larger applications (i.e., up to 87 billion of instructions) shows a better OVPsim performance in all configurations, while the gem5 atomic remains stable. With four host cores, the longest workload achieves up to 3910.86 MIPS when deploying the OVPsim-FIM. In a similar scenario, the gem5 atomic achieves 12.52 MIPS, approximately 325 times faster. The OVPsim speed increases as the application grows due to the just-in-time engine algorithm, and thus benefiting from larger applications. For example,

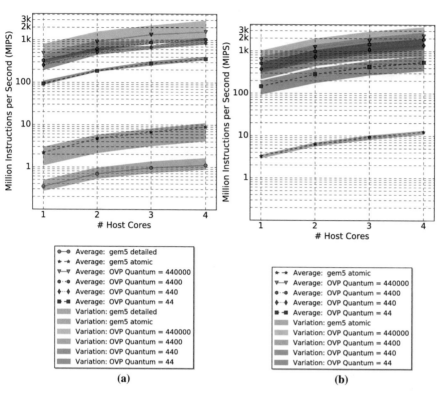

Fig. 4.13 Multiple VP-FIMs simulation speed and scalability using a quad-core host processor in terms of MIPS. (**a**) Rodinia benchmarks. (**b**) NPB applications

by comparing the larger and smaller applications, the simulation speed ranges from 1190 to 3910 MIPS (i.e., a 3.28 increase) where the gem5-FIM atomic difference is less than 20%, varying from 12 to 10 MIPS.

4.3 Closing Remarks

This chapter presented three additional contributions. First, Sect. 4.2 compares the accuracy between instruction-accurate (i.e., OVPsim) and cycle-accurate (i.e., gem5) virtual platforms. Further, Sect. 4.2.2 explores the relationship between the instruction-accurate engine and mismatch with the gem5. Performance is fundamental to early design space explorations, for this purpose this work shows the OVPsim-FIM peak of simulation speed around 4000 MIPS considering a quad-core host processor, Sect. 4.2.3. The OVPsim improves the simulation performance with larger workloads in the range of hundreds of billions of instructions, in contrast, both gem5-FIM atomic and detailed do not show any considerable speed variation by varying the workload size.

Chapter 5
Extensive Soft Error Evaluation

This chapter utilizes the power of the fault injection framework previously detailed to investigate soft error reliability from two distinct approaches: (1) the impact of different software stacks including OS, parallelization library, ISA using more than 50 distinct embedded and high-performance applications with up to 85 billion of object code instructions (Sect. 5.1); (2) the benefits of the novel fault injection techniques and error inspection described in Sect. 5.2.

5.1 Soft Error Evaluation Considering Multicore Design Metrics/Decisions

The emerging use of multicore processors requires specialized libraries, in this way, including an additional complexity to the system reliability assessment. Section 5.1.2 extensively explores the use of OpenMP- and MPI-based applications reliability. Different ISAs are available during early design space explorations, for example, ARMv7 32 bits and ARMv8 64 bits. In this context, Sect. 5.1.1 investigates the impact of such distinct ISAs on the system behavior under fault injection.

5.1.1 ISA Reliability Assessment

5.1.1.1 Execution Time and Workload

The ARMv7 workload for a single faultless execution has an instruction count that ranges from 299 million to 87 billion, with an average of 16 billion of instructions. In contrast, the 64-bit architecture applications execute in average

F. Rocha da Rosa et al., *Soft Error Reliability Using Virtual Platforms*,
https://doi.org/10.1007/978-3-030-55704-1_5

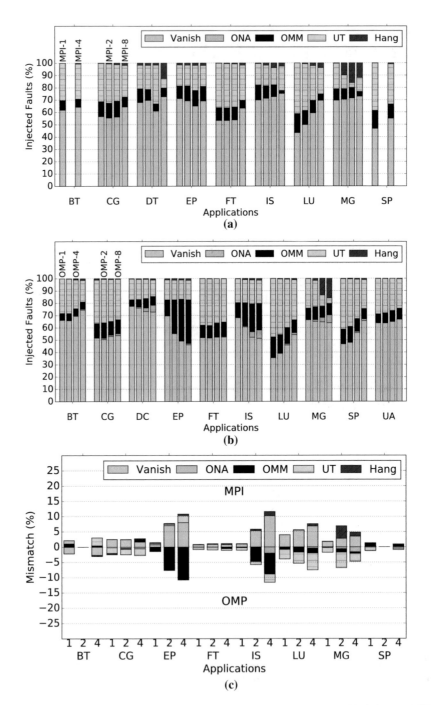

Fig. 5.1 Fault injections using a multicore ARM Cortex-A72 processor (ARMv8). (**a**) MPI benchmarks. (**b**) OpenMP benchmarks. (**c**) Mismatch

Table 5.1 NPB workload summary

Description		Minimum	Average	Maximum
Executed instructions	ARMv8	41.1×10^6	654×10^6	3.08×10^9
	ARMv7	299×10^6	16.5×10^9	87.4×10^9
Simulation time single run (s)	ARMv8	35	437	2134
	ARMv7	163	7929	42,763
Single campaign run (h)	ARMv8	77	971	4742
	ARMv7	363	17,620	95,028
Total fault campaign (h)				
ARMv8	82,820	ARMv7		1,152,160

654 million instructions, varying from 41 million to 3 billion. Table 5.1 summarizes the workload regarding simulation time and the number of executed instructions with average, smaller, and larger cases.

Applications executed using the ARMv8 ISA present a significant performance improvement when compared to the ARMv7. In some cases, the speedup reaches up to 10 times. This performance gain can be pinpointed to the removal of several legacies features (e.g., fast and multilevel interruptions, conditional instructions) and to significant improvements made in the floating-point (FP) unit by adding new specialized instructions and increasing the number of FP registers. The ARMv7 often resorts to the ARM software FP library to perform some operations and thus increasing execution time. This choice was made automatically by the compiler. The evaluated workload employs HPC scientific applications with some of them heavily depending on FP computation, leading to a significant performance boost. The executed instruction count for each application where the average value reduces from 16 billion (ARMv7) to 654 million (ARMv8) instructions (Table 5.1). A shorter execution time improves the ARMv8 mean time between failures (MTBF) as it has a smaller probability of being stroke by a radiation event for a given particle fluence.

5.1.1.2 Register File Size

The new 64-bit ISA also enlarges the integer register file, from 16 to 33 registers, increasing the number of possible targets for fault injection by a factor of four. However, the compiler algorithm uses a reduced subset of the available registers for critical operations (i.e., load/store and control flow) leaving other registers for variables or unused. As in this experiment, each register suffers an identical number of fault injections, critical registers (e.g., program counter, stack pointer, those used on load/store and control flow operations) are less likely to face faults in the ARMv8 rather than in the ARMv7.

5.1.1.3 Branches and Function Calls

The *Hang* error occurs when the target application control flow is severely affected by transient faults, in most cases, leaving the algorithm in an infinite loop. Analyzing individual parameters not always expose direct relationships between profiling data and fault injection campaigns. For example, the mean branch composition from the total executed instructions is 19.24% ($\sigma = 0.21$), 14.08% ($\sigma = 0.56$), 17.65% ($\sigma = 0.03$), and 12.01% ($\sigma = 0.36$) considering the four macro scenarios MPI V7, OpenMP V7, MPI V8, and OpenMP V8, where σ is the standard deviation. While the ARMv8 displays a 2% decrease in the mean branch occurrence compared with the 32-bit architecture, the application behavior under fault influence does not show any meaningful impact. Additionally, *function calls* variation also does not display any distinctive link with Hangs incidence. By combining both figures, nonetheless, it is possible to uncover a correlation between this new index value (i.e., number of function calls times number of branches) with the Hang incidence after comparing the 130 scenarios. Table 5.2 exemplifies this behavior using the *IS* application as a case study, note that this new index value and the Hang percentage increase simultaneously, an observable behavior through several scenarios. The ARM ISA (i.e., ARMv7 and ARMv8) use distinct instructions to compare the conditional statement (e.g., cmp) and another to perform the control flow branching, while function calls use unconditional branches (e.g., jumps) in conjunction with argument registers.

Table 5.2 Hang occurrence compared with the normalized function calls multiplied branches (F*B)

Scenario	Parameter	Number of cores		
		Single	Dual	Quad
IS MPI V7	Hang (%)	0.413	0.625	3.000
	Branches	56.0×10^6	58.0×10^6	196×10^6
	F. Calls	22.6×10^6	23.1×10^6	26.9×10^6
	Index F*B	1.000	1.024	1.700
IS OMP V7	Hang (%)	0.288	0.313	0.400
	Branches	54.1×10^6	54.3×10^6	54.7×10^6
	F. Calls	21.7×10^6	21.7×10^6	21.7×10^6
	Index F*B	1.000	1.001	1.002
IS MPI V8	Hang (%)	0.438	1.850	3.800
	Branches	11.2×10^6	15.9×10^6	17.6×10^6
	F. Calls	2.85×10^6	3.35×10^6	4.84×10^6
	Index F*B	1.000	1.302	1.799
IS OMP V8	Hang (%)	0.225	0.925	1.175
	Branches	7.99×10^6	9.05×10^6	9.50×10^6
	F. Calls	1.81×10^6	2.05×10^6	2.06×10^6
	Index F*B	1.000	1.172	1.194

5.1.1.4 Memory Transactions

UTs (i.e., unexpected terminations) originate from OS *segmentation fault* exceptions, which means that the program has attempted to access an area of memory outside its permissions. At instruction level, the address generation of memory access operations (e.g., load and stores) is compromised by transient faults in the source registers to lead to wrong address calculations. The reduced number of ARMv7 registers to perform address calculations leads to the use of load/store templates by the compiler to diminish the computational cost of register recycling. In other words, the ARMv7 compiler continuously utilizes the same register to perform memory transactions (e.g., R0–3 and SP). As consequence of this behavior, increasing the number load/store operations can lead to a more significant UT occurrence in the target application using an OS on top of the ARMv7 processor. Table 5.3 shows the soft error results (e.g., Vanish, UT, Hangs) alongside the memory access figures for some examples of the behavior mentioned above. By increasing the percentage of memory transactions (i.e., load and stores instructions) in applications such as MG and IS increases the UT ratio. For example, MG application memory-oriented operations for single- and quad-core processors are 15, and 22% while the UT occurrence increases from 22 to 30%. Further, increasing the core count alone does not reduce the UT percentage as is possible to note by comparing scenarios (1, Table 5.3) against (2) where both have similar memory instruction occurrence.

The 64-bit architecture exhibits a similar behavior considering FP memory transactions, supporting the claim above that wrong address calculation related to memory access, as FP instructions are exclusively used for computation and not for control flow operations (e.g., branches and jumps). Table 5.4 displays nine scenarios (A–I) of soft error analysis and FP memory figures. Reducing the memory transactions participation from the total number of executed instructions for LU (A–C) and SP (D–F) applications shows a UT occurrence reduction trend. Scenarios (G–I) reinforce this hypothesis by demonstrating that a constant memory-oriented instruction incidence leads to a regular UT percentage.

Table 5.3 ARMv7 memory transactions and soft error classification for selected scenarios

	Scenario	Vanish +OMM +ONA	UT	Mem. inst. (%)	RD/WR ratio
1	MG MPIx1	78	22	15.8	1.18
2	MG MPIx2	78	22	16.3	1.12
3	MG MPIx4	70	30	22.5	2.83
4	IS MPIx1	80	20	18.0	0.85
5	IS MPIx2	80	20	19.0	0.83
6	IS MPIx4	70	31	26.0	2.73

Table 5.4 ARMv8 memory transactions and soft error classification for selected scenarios

	Scenario	Vanish +OMM +ONA	UT	Mem. inst. (%)	RD/WR ratio
A	LU OMPx1	57	48	29	1.9
B	LU OMPx2	59	45	27	1.9
C	LU OMPx4	67	40	22	1.9
D	SP OMPx1	57	42	35.1	1.5
E	SP OMPx2	59	40	34.0	1.5
F	SP OMPx4	70	32	28.5	1.5
G	FT MPIx1	62	37	25.7	1.00
H	FT MPIx2	62	37	24.6	0.95
I	FT MPIx4	62	36	23.7	0.95

5.1.2 Parallelization API

The OpenMP library uses a series of the fork and joins approach to parallelize loop statements (e.g., *for*, *while*) where the API automatically creates children threads, being suitable for shared memory. In contrast, the MPI standard is adequate to distribute systems due to the use of a message-oriented parallelization technique, which requires the direct user parallelization regarding thread creation and communication. Figures 5.2 and 5.3 display fault injection campaigns and mismatches comparing the MPI and OpenMP applications.

5.1.2.1 Serial vs. APIs

When we compare the serial implementation with either parallelization libraries on both architectures, some patterns can be observed. In ARMv7 MPI, only CG has a small improvement in the number of UTs, while in IS and MG the number of UTs and Hangs increases. Considering the OpenMP versions, no significant variation can be found. For the 64-bit application set, *CG, LU, MG, SP, and UA* the number of UTs diminishes. Further, CG application maintains the number of UTs when the number of cores increases. The same cannot be said about the other application, where the number of UTs diminishes with the increase in core count. Other applications have negligible variations.

5.1.2.2 Vulnerability Window

Within a software stack, some components are more critical to the system correct behavior. For example, targeting a thread scheduling function with faults has a potentially more hazardous effect on the system reliability than a purely arithmetic code portion. By comparing these critical functions active periods against application execution time, it is possible to define a time interval called *vulnerability*

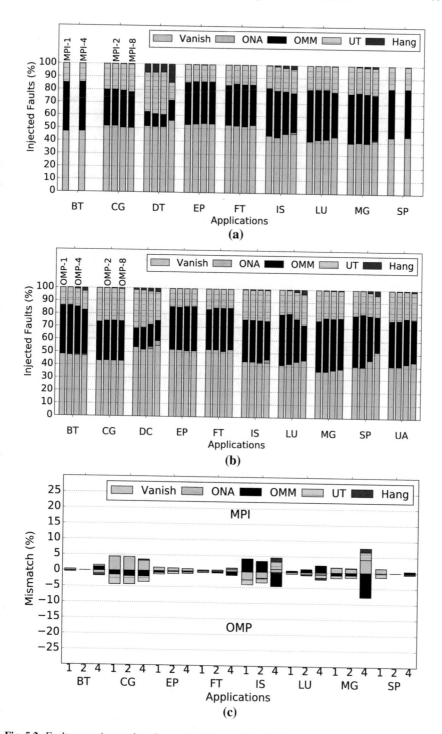

Fig. 5.2 Fault campaigns using the gem5-FIM targeting distinct parallelization paradigms. (**a**) gem5 atomic using MPI applications. (**b**) gem5 atomic using OpenMP applications. (**c**) gem5 atomic mismatch MPI vs. OpenMP

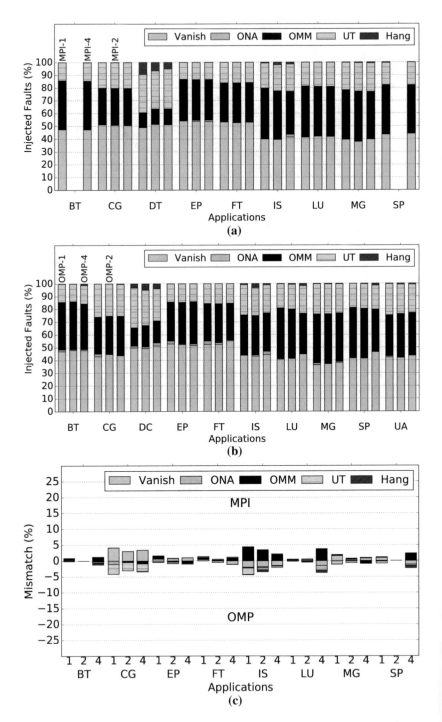

Fig. 5.3 Fault campaigns using the OVPsim-FIM targeting distinct parallelization paradigms. (**a**) OVPsim using MPI applications. (**b**) OVPsim using OpenMP applications. (**c**) OVPsim mismatch MPI vs. OpenMP

window, which varies with the number of calls and executions of the function. Using the NBP benchmark suite provides a real high-performance workload, enabling a more accurate evaluation of the OpenMP and MPI libraries impact on the system reliability. Due to its reduced vulnerability window, the parallelization mechanism has a limited effect on the final reliability assessment, less than 23% in the worst case.

From the 44 possible comparisons between the MPI and OpenMP scenarios, in 38 the MPI has a higher masking rate (i.e., executions without any errors) due to two main reasons: *First*, MPI applications have a better workload balance among the used cores, in other words, the number of executed instructions per core is very similar. For instance, the average difference concerning executed instructions per core is around 4% for both ARMv7 and ARMv8 considering MPI applications, while the OpenMP variation reaches up to 16%. As the OpenMP does not fully utilize the available cores due to the fork/join parallelization approach where a loop statement executes in parallel and other code portions hastily. By contrast, the MPI has individual and independent working threads for each running core providing a better workload balance during its execution. Whenever a core is sub-utilized, it executes a thread scheduling policy and when no thread is suitable the core waits in a sleep mode. By consequence the kernel probability to suffer a transient fault increases, as the scheduling is more often executed. *Second*, OpenMP benchmarks have a smaller execution time, 16% on average, compared against the MPI applications. By consequence, diminishing the vulnerability window of the MPI inner-functions when comparing against the OpenMP. Further, the longer execution increases the chance of the injected fault being erased due to software and microarchitectural masking mechanisms.

5.2   Focused Fault Injection Results

The traditional fault injection flow focuses on error rate estimation for a hypothetical radiation fluence through a probabilistic figure. Nevertheless, it does not stress all possible fault injection outcomes satisfactorily. In other words, it exposes the average error rate, ignoring that application domains have different reliability constraints. For example, considering the scientific applications, a silent memory corruption (SDC) denotes a higher reliability issue than an unexpected termination (UT). A UT is easily detectable, and the HPC environments already offer solutions as periodical checkpoints and restartable jobs to deal with this issue. In contrast, SDC detection or correction techniques are computationally costly [27] and the incorrect detection may affect the final application result [11]. The incorrect SDC treatment can jeopardize the scientific investigation. In the automotive domain, an unexpected termination leads to harmful consequences [110], and in contrast, SDCs are mostly benign in a real-time application.

The set of experiments presented in this section aims to demonstrate the OVPsim-FIM extension usability during early design space explorations stages concerning

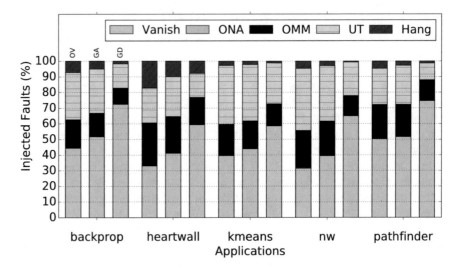

Fig. 5.4 Soft error classification according to the system behavior under register fault injection considering 5 Rodinia benchmarks executing onto single-core ARM Cortex-A9, regarding the three fault injection module (FIM) implementations: gem5-FIM atomic (GA), gem5-FIM detailed (GD), and OVPsim-FIM (OV)

the software reliability. The target architecture comprises an ARM Cortex-A9 interconnect through a bus on a dedicated one-gigabyte memory. The first experiment goal is to provide a reference for register targeting faults in comparison with the promoted novel fault injections techniques. It software stack included a Linux 3.13 kernel and selected Rodinia OpenMP applications (i.e., backprop, heartwall, kmeans, nw, and pathfinder) both compiled with the cross-compiler *arm-linux-gnueabi-gcc 6.2.0 20161005*. The following experiment, Fig. 5.4, assesses the selected benchmarks reliability under an 8000 register file fault injections aiming to estimate the percentage of errors that are not masked by each benchmark. Further, this setup includes three different VP-FIMs: the OVPsim-FIM (OV), gem5-FIM atomic (GA), and gem5-FIM detailed (GD). As shown in previous sections, the gem5 detailed presents more vanished faults due to the masking mechanisms from the microarchitecture representation.

The second case study targets the application virtual memory primarily and to demonstrate the OVPsim-FIM extension necessity and to establish comparative results; it also shows the outcome of a physical memory fault injection campaign. Figure 5.5 displays the selected Rodinia applications in five different fault injection scenarios: 80,000 physical memory faults considering the (1) gem5-FIM (PHY-GA) atomic and the (2) OVPsim-FIM (PHY-OV), and three 8000 faults targeting the application (3) entire VA space (VA-ALL), (4) only data sections (VA-DATA), and (5) the code section (VA-CODE). Physical memory fault injection campaigns faith-

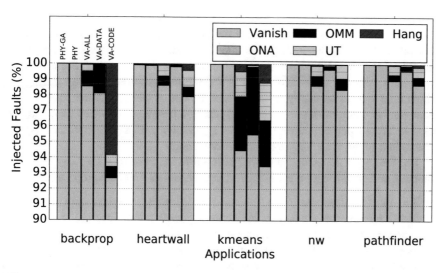

Fig. 5.5 Memory-based fault injection campaigns targeting physical memory gem5-FIM atomic (PHY-GA) and OVPsim-FIM (PHY-OV), comparing with OVPsim-FIM extension VA the entire range (VA-ALL), the code section (VA-CODE), and data section (VA-DATA)

fully represent the memory behavior under the influence of soft errors. However, the extracted information has little or none relevance to understand the application reliability. To achieve meaningfully statistical figures the experimental setup was lengthened by ten times when compared with the virtual memory experiments; where 99.9% of the total injected fault. This small physical memory scenario (i.e., 80,000 faults for five applications) requires approximately 4400 simulation-hours, and after this long simulation, only 200 from 80,000 fault injections do not present a masked outcome in average. The one-gigabyte memory has 137,438,953,472 available bits multiplied by the execution time in which each intersection is a possible target. At any time, the application accesses a limited memory range (i.e., few kilobytes) during its execution, even the data-intensive algorithms process small memory blocks per access. Consequently, the chance of physical memory bit-flip impacting the application behavior is minimal; also, this event timing should be precise and occur before reading access.

The virtual address fault injection technique reduces the target range, thus focusing on the application code and data. Faults targeting the code segment exclusively (i.e., VA-CODE) exhibit a larger occurrence of control flow related errors (i.e., UT and Hangs) due to the instruction representation changes. As a possible utilization of this technique, several mitigation techniques replicate or introduce a new instruction in the original application code after assuming a fault-free code. With this promoted fault injection technique it is possible to explore the SWIFT-based mitigation methods and improve its coverage by adding a new approach angle. In contrast, the data section faults incur in memory corruptions, and reduced number of control flow errors as the applications data structures are affected. Even when reducing the memory range from the full memory to the virtual memory range a significant portion of the fault injections still leads to a vanished outcome. The compiler includes hundreds of functions and data structures from adjacent Linux and C libraries alongside the own application source. Some of them only run during the application startup, and the majority never execute during the simulation and consequently, a large part of the virtual memory, either code or data, does not have any influence over the application behavior. This technique may be used to target a particular process (i.e., an application with a distinct virtual address space) in a sophisticated software stack, for example.

The fault injection scope is further reduced in the third case study by using the function code and lifespan techniques which use targets four backprop functions:

1. The most timing-consuming function (bpnn_adjust_weights._omp_fn.1), this function represents the application kernel OpenMP parallelization;
2. A sequential portion of the application (bpnn_layerforward);
3. The OpenMP synchronization barrier (gomp_barrier_wait_end);
4. The Linux kernel next processes selection algorithm (pick_next_task_fair).

Figure 5.6 shows the four 8000-fault campaigns for both function code and lifespan (LF) techniques. The lifespan technique targets the 16 general-purpose registers (r0-r15) including the program counter (PC) and stack pointer (SP) assigning each one 500 faults from the total 8000. The most time-consuming function bpnn_adjust_weights._omp_fn.1 is the application kernel parallel phase and data intensive. This function behavior under both fault techniques (lifespan—LF—

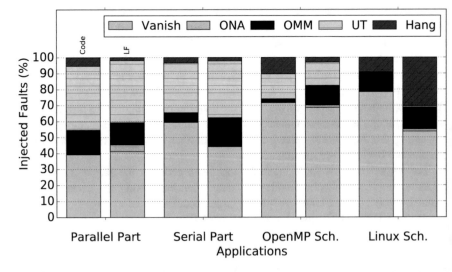

Fig. 5.6 Function-based fault injection techniques targeting four different functions *bpnn_-adjust_weights._omp_fn.1, bpnn_layerforward, gomp_barrier_wait_end, and pick_next_task_fair)* targeting both Code and Lifespan (LF)

and instruction code) resembles the compartment when targeting the complete application as it accounts for approximately 20% of the total execution time. The sequential function (2) has a smaller number of hangs due to it execution being limited to the application initialization, as the subsequent code execution masks most of the faults.

The OpenMP gomp_barrier_wait_end is a short (i.e., few lines) function that acts as a thread synchronization barrier. Therefore, the code and register fault injections have a similar behavior causing UT (i.e., Linux segmentation) errors due to the wrong address calculation. Code faults show a more significant number of hangs in the first three cases, as the small function code increases the probability of a control flow error leading to an infinite loop. The registers are continuously overwritten at each function invocation and in contrast, code faults remain incorrect for more extended periods. The Linux kernel function (4) has a complex chain of control flow statements (i.e., if and else). Whenever a fault strikes an addressing register (i.e., r3, r4, and r5 for this particular function), the control flow is severely affected leading to kernel malfunction and consequently, infinite loops (Hang) and therefore other registers, beside the PC and SP, have little or none impact. In several cases, it causes an SDC error as the target register remains in the function context stored in the stack. In contrast, the code targeting in this particular function shows as smaller impact, as large portions of the function are not executed due to the control flow statements.

5.2.1   Case Study

To demonstrate the soft error analysis capabilities of proposed fault injection techniques, we select a matrix multiplication (MM) kernel as a case study due to its application in several fields and branches of science. During this case study, we will subject the MM to the fault injection techniques presented in Sect. 3.7, targeting distinct software components alongside a customized error classification module. Each technique covers different aspects of the MM considering its variables and critical functions in an isolated manner, demonstrating the importance of providing software engines with appropriate means that can lead to application reliability improvements. Experimental setup comprises an ARM Cortex-A9 processor, executing Linux kernel (3.13), and the MM kernel using 300-wide 32 bits integer square matrices as inputs and output.

5.2.1.1   Sequential and Parallel MM

The first fault injection campaign deploys the MM kernel in two versions: (1) sequential implementation that uses a simple iteration to perform the multiplication, and (2) a parallel MM using the Pthreads library to create two working threads. Figure 5.7 shows the two MM implementations subjected to 8 fault campaigns of

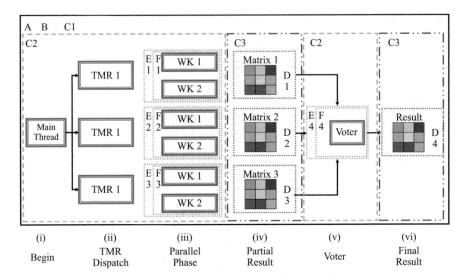

Fig. 5.7 TMR-Based Matrix Multiplication execution flow and fault target locations according to Table 5.5. WK refers to working thread

8000 fault injections each (i.e., totaling 128,00 simulations), considering the six fault injection techniques:

1. random registers (A),
2. physical memory (B),
3. virtual memory (VM) entire range (C1),
4. VM code section (C2),
5. VM data sections (C3),
6. result matrix (D1),
7. multiplication function object code (E1),
8. multiplication function lifespan (F1).

The sequential MM presents a larger OMM (e.g., silent data corruption) due to the high number of memory reads and writes with a small number of operations for each matrix cell. The resulting scenario leads to a significant number of silent data corruptions in the target system memory as a consequence of dirt registers used for the MM. Underlying implementations are susceptive to UTs (i.e., unexpected terminations) due to the incorrect memory address computation caused by registers under fault influence, which may lead to errors such as *segmentation fault*. However, the parallel MM presents a substantially more significant number of UT when compared to the sequential version. This behavior is explained because Pthreads scheduling algorithm increases the application control flow complexity, which might incur in more wrong address computation during the MM execution.

5.2.1.2 Triple Modular Redundancy

As identified in the previous section, the injection of bit-flips severely impacts the matrix multiplication kernel operation due to its simplicity and small code. Aiming to reduce the amount of detected silent data corruptions, we propose the use of the MM-TMR to improve its reliability by using a well-known fault mitigation technique the *Triple Modular Redundancy* (TMR). This technique adopts three independent *parallel MM* instances (i.e., six working threads in parallel) enabling one incorrect execution to be masked by a voting process at the end of MM execution (i.e., a function vote the majority from the partial results). The TMR version corrects most of the errors originated from the input and output matrices. Nevertheless, the TMR implementation using the Pthread library increases the occurrence of UT.

To improve the experiment, we included a custom soft error analysis step, which compares the four matrices (i.e., each TMR replica and the voter) alongside two additional error classifications considering three possible outcomes:

1. All matrices are identical, in this case, the SOFIA classifies the error according to one of the five default classes (e.g., Vanish, UT, Hang). Note that OMM and ONA only occur if the result matrix is correct, and thus being considered benign errors in this context.
2. If one TMR matrix does not match the other replicas, the voter will mask the error and produce the correct result. Nevertheless, in this case, the simulation diverges concerning the number of executed instructions from the faultless run, which leads to a false-positive error (i.e., control flow error with incorrect memory) in traditional fault injection flows. The SOFIA classifies this execution context as **Corrected** to signal the appropriate behavior, (i.e., even with the context mismatching the reference execution of the final matrix is correct).
3. The third possible outcome originates from an incorrect voter execution (i.e., the three TMR matrices are identical and differ from the voter matrix) due to the fault injection being classified as **Voter Error**.

Table 5.5 describes 17 distinct fault injection scenarios targeting the MM-TMR, while Fig. 5.8a shows the results considering a single- and quad-core ARM Cortex-A9 processor where each fault injection scenario comprises 8000 faults. Register-based fault injection (A, E, and F) displays a considerable amount of UT (i.e., Linux OS segmentation faults in this context), around 40% due to the wrong address computation using registers under fault influence. In contrast, the memory-based technique errors depend on the stroke region, for example, targeting the 1 Gb physical memory (using technique B) would result to a minimal number of errors (i.e., masking rate of 99.95%) as the benchmark accesses a limited memory range (i.e., few dozen kilobytes). The complete VM range (C1) and data sections (C3) present a similar behavior as most of the faults hit the application 300-wide square matrices due to its size (i.e., each one possessing 360 kilobytes or 20% of application size). The code section (C2) contains, besides the application code, hundreds of Linux and C libraries unused functions added by the compiler, leading to higher

Table 5.5 Deployed fault injection techniques into the TMR-based matrix multiplication reliability exploration

#	Target	#	Target
A	Register file	E1	Function code first replication
B	Physical memory	E2	Function code second replication
C1	VA complete	E3	Function code third replication
C2	VA code section	E4	Function code voter
C3	VA Data sections	F1	Function lifespan first replication
D1	Variable matrix 1	F2	Function lifespan second replication
D2	Variable matrix 2	F3	Function lifespan third replication
D3	Variable matrix 3	F4	Function lifespan voter
D4	Variable matrix result		

masking rate. By individually targeting the matrix replicas (D1–3) we exercise the TMR main functionality resulting in an almost complete error coverage. The fault campaign D4 leads to a 99.9% masking rate as the final result is composed of the voter function at the application end, which incurs in a narrow sensitive window (i.e., any faults previously present in this matrix are overwritten).

Passing the focus to function criticality, groups E (E1–4) targets the function object code, while the F (F1–4) injects bit-flips into the processor register file during the target function execution (i.e., the function lifespan). The matrix multiplication kernel has a compact code of few dozens of lines executing for extended periods, which composes most of the application execution time. Single- and quad-core processors show a similar rate of correct results (i.e., vanish, SDC, and corrected) when targeting the function object code (E1–4). However, their composition diverges while the single-core processor presents a more significant SDC rate (i.e., the MM result is correct with silent data corruptions on the memory) the multicore system displays a larger masking rate. Further, the multicore system reveals a higher number of *Hangs* due to the longer and more significant executions of the PTHREAD scheduling policy leading to unrecoverable control flow. Random register (A) and Lifespan (F1-3) techniques show similar behaviors under fault injection as the MM application spends 95% on those multiplication functions. Directly targeting the voter function shows a behavior not seen when targeting the complete application with random faults due to its short execution time, and thus, demonstrating the necessity of more detailed fault injection framework. Subjecting the voter code (E4) and lifespan (F4) to fault injection causes an erroneous matrix voting, which is a severe error in this context.

5.2.1.3 Improving the Triple Modular Redundancy

The initial MM-TMR solution provides complete coverage to fault injections for the replicated data (i.e., the partial matrices) while the control flows still prone to unexpected terminations. By using the promoted framework, it is possible to pinpoint

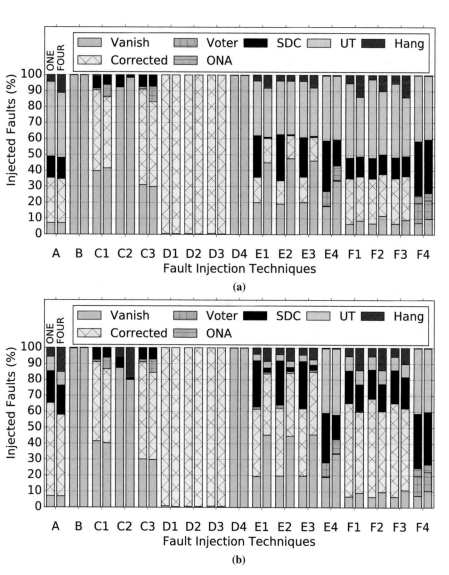

Fig. 5.8 MM soft error vulnerability analysis considering a single- and quad-core ARM Cortex-A9 processor. (**a**) MM-TMR. (**b**) Improved MM-TMR

the major UT cause as OS segmentation faults in one of the thread replicas that terminates the complete application even if the other replicas had not experienced any errors. To mitigate this issue, we modified the application algorithm to include a segmentation handler for each replication, and consequently, the improved MM-TMR (MM-TMR-I) finishes correctly even if one of the replicas generates an OS segmentation fault. The experiments displayed in Fig. 5.8b reproduce the 17 fault

injection scenarios mentioned above for the MM-TMR-I version using the single-
and quad-core processors. The MM-TMR-I improves the MM kernel reliability by
achieving of up to 90% of coverage (i.e., with correct final results) in contrast to the
50% of the traditional TMR considering register-based fault injections targeting the
replicas working threads. Fault injection techniques (D4, E4, and F4) targeting the
voter function and data remain unchanged without any modification being made in
its code.

5.3   Closing Remarks

The chapter demonstrates the promoted fault injection framework to evaluate
distinct design decisions during the initial development phase. This work explores
the novel 64-bit ARM architecture and assesses its reliability under a soft error in
Sect. 5.1.1. The larger register file from the new ISA increases the masking rate as
the application fails in taking advantage of the extra registers. Further, this work
(Sect. 5.2) investigates the impact of parallelization APIs in the overall software
stack reliability by comparing serial, OpenMP, and MPI implementations of the
same benchmarks. Fault campaigns show a smaller incidence of the parallelization
API in the overall system reliability due to the limited time ratio of those libraries
in comparison with the total execution time.

The third contribution involves the validation/use of proposed novel fault injections techniques and error analysis methodologies. In this Sect. 5.2 we demonstrate
the adaptability to different aspects of the fault injections campaigns scenarios. With
the proposed framework, software engineers can focus on the resiliency exploration
to specific software stack components.

Chapter 6
Machine Learning Applied to Soft Error Assessment in Multicore Systems

While commercial multicore processors based on 10 nm process node are already available, processing components manufactured in 5 nm are expected to be released in the market by, approximately, the end of the decade. Such processors are likely to be integrated into many electronic computing systems from a diverse range of industrial sectors, including medical, automotive, and high-performance computing (HPC). The increasing number of internal elements (e.g., cores, memory cells), coupled with the high clock frequency operation of multicore processors is making the aforementioned systems more vulnerable to radiation-induced soft errors [99]. The occurrence of soft errors can lead to failures of critical parts of a system, which might ultimately incur in financial or human life losses. Thus, assessing and mitigating the occurrence of soft errors in such systems is critical to accomplish a reliable and efficient operation.

The increasing software and hardware complexity of such systems imposes exploration challenges, including: (*ch1*) conduct a large number of fault injection campaigns within a reasonable time; (*ch2*) provide engineers with detailed observation of system's behavior in the presence of faults; and (*ch3*) identify relationships or associations between application profiling and specific platform parameters in large data sets resulting from the fault campaigns. Aiming to overcome the challenges *ch1* and *ch2*, researchers are incorporating fault injection capabilities into virtual platform (VP) frameworks [27, 28, 45, 49, 61, 92, 101], enabling the detailed observation and analysis of complex software stacks and multicore architectures under the presence of faults at early design phases.

Machine learning has providing robust solutions for broad range of complex problems, such as weather forecast or oil exploration. For instance, Video-on-demand distribution algorithms employed by Netflix, Amazon, or YouTube recommend films and series based on users interests. State-of-the-art oncologists are using machine learning to uncover early breast cancer symptoms by analyzing millions of image [65]. Using this extensive set of images the algorithm is capable of finding complex patterns, which otherwise would be impossible to humans. The main

F. Rocha da Rosa et al., *Soft Error Reliability Using Virtual Platforms*, https://doi.org/10.1007/978-3-030-55704-1_6

contribution of this chapter relies on the exploration of supervised and unsupervised machine learning techniques that can be used to identify the correlation between fault injection results and application and platform microarchitectural characteristics.

6.1   Machine Learning

Searching for another approach computer scientists started investigating how to give computers the ability to learn without being explicitly programmed [91]. Machine learning (ML) is the field of study that uses the target system past (knowledge) to predict its future behavior (intelligence). In other words, machine learning is an algorithm or model capable of learning from a collection of inputs without requiring rules-based programming (i.e., if-else)[1, 102] by transforming this information into actions or predictions [102].

Machine learning algorithms can be categorized into three broad groups:

- **Supervised Learning** goal is to predict the effect of one set of observations (i.e., input, features, attributes) has on another dependent variable (i.e., output, label, class). In other words, supervised learning algorithms make predictions based on a set of training examples using two main approaches: A regression technique models the target system using mathematical equations producing a continuous value output (e.g., linear regression) to approximate the target system behavior. Classification algorithms divide a data set into smaller subsets by evaluating its attributes (e.g., decision tree).
- **Unsupervised Learning** searches for patterns on unlabeled datasets (i.e., without a dependent variable output). While supervised learning correlates two groups of observations, unsupervised learning describes the system features into more abstract levels of representations. For example, K-means is a well-known example of an unsupervised learning algorithm to subdivide n observations into k clusters where each observation belongs to the cluster with the nearest mean.
- **Reinforced Learning** gains experience (i.e., knowledge) through a series of trial-and-error training sessions where a cost function calculates reward or punishment value depending on its prediction. By minimizing the target cost function, the reinforced learning algorithm improves the system prediction until reaching a predefined quality threshold.

6.2   Machine Learning Applied to System Reliability

Machine learning has been employed in different domains in recent years to recognize patterns and predict the future system behavior. For example, [50] trains a random forest ML algorithm with multiple TCAD simulations to estimate the

Table 6.1 Most recognizable virtual platform fault injection simulators

Year	Author	Machine learning	Features
2015	[6]	Supervised learning	Linear regression
2016	[30]	Supervised learning	Decision tree
2016	[104]	Supervised unsupervised learning	SVM, k-nearest neighbors, decision tree, and other four algorithms
2017	[100]	Supervised learning	Linear regression, SVM, k-nearest neighbors, decision tree, and Ada boost regressor
2017	[76]	Supervised learning	Artificial neural network

SER of an SRAM Cells. Giurgiu et al. [42] use field information as a training set for a random forest algorithm to predict a DRAM number of errors. Considering the challenges involved in soft error assessment, some researchers are studying the applicability of ML techniques to speed up its simulation, prediction, or mitigation. Table 6.1 shows a review of the state of the art on soft error investigation on multi/manycore system using machine learning approaches.

Vishnu et al. [104] analyzes the impact of multi-bit memory errors targeting both permanent and transient faults on large-scale applications. The training sets consist of fault injections targeting the applications primary data structures. This work tests eight different ML algorithms and compares their predictions (i.e., the application error probability) with the ground truth (fault injection). This paper shows an average error detection of 90% considering three applications under the influence of permanent and transient faults. The fault injection and analysis are tightly coupled with the application, requiring an excellent understanding of the application execution.

The work [100] proposes an online adaptive SDC detection algorithm using machine learning. First, it investigates different supervised ML algorithm execution time to select the best suitable for an online detector. The proposed SDC detector runs after each application iteration, applying one ML algorithm (from five), predicting the error impact on the system. The system has a high percentage of correct predictions, 99%, over eleven HPC applications.

Nie et al. [76] collect several parameters (e.g., power, GPU error logs, and temperature) of a supercomputer during four months. This work correlates these pieces of information for each node searching for error patterns hidden in this big data problem. After an initial statistical analysis, this work employs an artificial neural network to predict the GPU error probability in different conditions.

Ashraf et al. [6] studies the propagation of transient errors on large-scale MPI application (i.e., up to 1000 cores). This work introduces additional instructions in the application code using the LLVM Intermediate Representation (IR) to inject faults and check for errors. Further, this instrumentation tracks the error through several MPI processes by monitoring communication messages, load–store operations, and function calls. With this information, the authors create fault propagation models to estimate the number of corrupted memory locations.

DiTomaso et al. [30] measures several NoC parameters such as temperature and wearing to feed an error prediction model called VARIUS. The VARIUS provides the error probability for different types of faults depending on the input parameters. All this information is supplied to a decision tree which predicts on-the-fly if a specific link will suffer *No Error*, *Few Errors*, or *Many Errors* before any package transmission. The proposed system adjusts the packet mitigation technique (i.e., CRC. SECDED, or retransmission) according to the model output. This mitigation technique requires a 2-bit CRC, 4-bit SECDED the decision tree, and control modules in every system router.

6.3   Problem Description

Soft errors reliability assessment is a time-consuming process which requires extensive fault injection campaigns, in some cases, taking thousands of simulation hours. Accelerating the assessment of soft errors impact during early design space explorations (DSE), particularity, in the software stack, is one these book goals. For this purpose, it was selected two virtual platforms to provide a flexible and fast fault injection framework, adequate for such explorations. Nevertheless, even virtual platforms cannot provide adequate fault coverage when executing more realistic workloads. For example, this work longest simulation (i.e., EP) takes 12 h of machine for a single run which emulates only 50 seconds of real time. Note that this simulation targets a single-core ARMv7 processor, memory, and cache using the gem5 atomic mode. During early DSEs, the target application undergoes several cycles of reliability optimizations to meet the desired system constraints. How can we reduce the number of fault injections (i.e., computational cost) and aid the software engineer to improve the system reliability?

6.4   Proposed Solution

Machine learning techniques are being employed to predict and model systems behavior considering multi-parameter optimizations. However, the reviewed works target distribute HPC applications using the ML approach to overcome the inherent overhead from such scenarios simulation. This work proposes the utilization of multiple machine learning techniques to improve multicore systems software stack design during early explorations. Figure 6.1 shows the traditional application development cycle without any modifications (a, b, d, and e) and the proposed solution. The software engineer describes the application and system in the first step (a), e.g., architecture, compiler, libraries, # of cores. The second phase (Fig. 6.1b) displays the traditional soft error vulnerability compressing a random fault injection campaign. Any fault injection flow can be adapted in the promoted solution, such as the ones based on FPGAs, VHDL simulation, and virtual platforms. For this

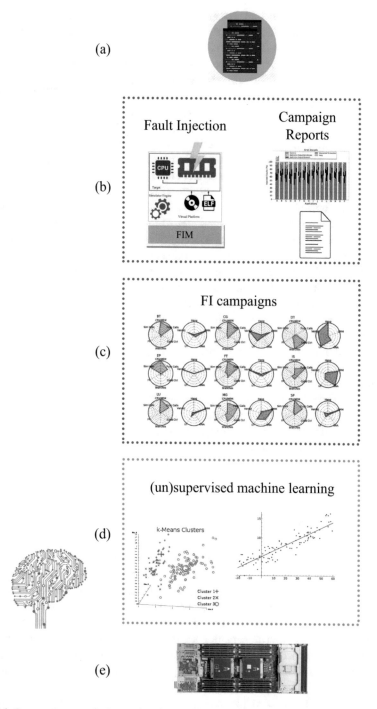

Fig. 6.1 Proposed system design cycle using machine learning. (**a**) Application design. (**b**) Soft error assessment. (**c**) Characterization. (**d**) Error symptoms. (**e**) Target system

work, we explore cycle and instruction-accurate virtual platforms in this role. Using this information, the software engineer recommends a mitigation technique or application modification (Fig. 6.1e) according to the target system (f) requirements.

The proposed solution adds two new development phases in the original flow (Fig. 6.1a, b, e, f): Characterization (c) and error symptoms detection (d). The characterization phase facilitates the access to raw gem5 microarchitectural parameters for an initial exploration. Software engineers can interactively search for parameters of interest (e.g., # of loads and stores) and compare with previous iterations or workloads. The application characteristics vary during the development cycle by algorithms or system modifications. This phase enables the user to evaluate the software adjustments impact on the system reliability over multiple iterations.

This work proposes an exploratory flow (Fig. 6.1d) to find the soft error correlations with multiple application characteristics. This flow applies supervised and unsupervised machine learning techniques to investigate the correlations between the fault injection results (i.e., vanish, hang, ONA, OMM, UT) and the application characteristics (e.g., cache statics, # of branches) without the presence of faults. This phase goal is to reduce soft error assessment time during early design stages by examining the impact of these characteristics on the application reliability. This work proposes a fully automated and standalone tool capable of searching and identifying the individual (or combinations of) parameters which present the most substantial relationship with each detected soft error. Initially, the investigation considers the microarchitectural information provided by the gem5 simulator (e.g., memory usage, application instruction composition). This tool collects over one million single fault injection raw outcomes alongside gem5 simulation reports to create statistical figures (e.g., the percentage of Vanish, Hangs) for each one of the explored scenarios. In possession of this database, the proposed framework applies supervised and unsupervised learning techniques to produce a model of each feature impact on the system reliability. With this observation, the software engineer may choose to alter the application or optimize the system parameters.

6.5   The Promoted ML Investigation Tool

This section provides an overview of multiple ML-related techniques developed in the promoted tool. The tool provides a generic exploration framework using machine learning supervised and unsupervised techniques to highlight the most relevant features. Initially, this framework combines two data sources: the gem5 microarchitectural statistics and the soft error vulnerability. However, its interface enables adoption of other information sources according to the investigation goals and direction. Further, exploration flows and other techniques can be created or modified because of the tool data structure parametrization on intermediate steps.

The proposed framework was developed using Python, taking advantage of available ML frameworks, in particular, the Scikit-learn module. Why Scikit-learn? Since its release in 2007, Scikit-learn [82] has become the most widely-used, general-purpose, open-source machine learning modules that are popular in both industry and academia. Further, this work adopts the pandas [71] module which provides a data structure designed for large-scale explorations. Besides this two modules, several other libraries where employed, such as matplotlib, numpy, scipy, multiprocessing. Sections 6.5.1–6.5.3 display multiple techniques to acquire and process the features embedded in the proposed tool.

6.5.1 Feature Acquisition

Feature acquisition comprises the data preparation phase of any ML framework, where the relevant information should be extracted from the raw input files. The promoted framework collects features from different sources requiring precise knowledge of the file format. The tool splits each file line into sub-strings considering the format parameters arrangement, dropping unnecessary information to reduce the memory usage.

Machine learning algorithms demand large amounts of raw data requiring an optimize storage method to reduce the memory utilization and increase its execution performance. Pandas is a popular python module which provides robust, expressive, and flexible data structures for data science applications. For 2D problems, the Pandas supports a rectangular grid data structure called dataframes (DFs). Unlike a matrix, it supports hierarchical multilevel index enabling sophisticated data analysis and manipulation. Further, each cell holds an object pointer enabling the combination of different data types numeric, character, logical in the same structure. Dataframes can be merged, concatenated, divided, and manipulated in multiple ways facilitating the access to data subsets data.

The raw input is transferred to a single dataframe where columns represent features (i.e., soft errors and microarchitectural) while rows represent the fault injection scenarios. This dataframe has several missing cells (represented as NaN) because not all scenarios have the same microarchitectural elements (e.g., number of cores). The missing data are replaced by zeros when necessary to enable the correct algorithm execution. Another dataframe functionality is the possibility to export and import data structures from *comma-separated values* (CSV) files. The promoted framework enables pre-processing or fast-forwarding some exploration steps, reducing the investigation time.

6.5.2 Feature Transformation

Some features original representation maybe not suitable for ML frameworks. For example, some algorithm requires float-point or integer values or the target information may be scattered across several columns. In this context, feature transformation is a set of techniques which creates new features using the already existing ones.

6.5.2.1 Rescaling

Some machine learning techniques perform complex equations using high-dimension equations where the data magnitude impacts on the algorithm performance. For example, a parameter with a nominal value of 1000 could perform worst than the same data rescaled to 10. For this purpose, this work adopts a Min–Max Scalar method that transforms features by rescaling the features. In this work, we rescale all features to a range between 0 and 10, which yield the best performance. This range improves the average performance of the Scikit-learn methods due to the smaller range from 0 to 10 instead of 0 to billions.

6.5.2.2 Normalization

The normalization function is another popular resize adopted method which scales the input vector individually to the unit norm (vector length) or another value. Each resizing function impacts the target data differently (e.g., flatting the values), the proposed tool supports distinct methods according to the target technique.

6.5.2.3 Merging Similar Features

Multiple features may contain similar information, and merging them is possible to highlight this behavior. For instance, the gem5 provides microarchitectural information subdivided by core, and consequently, the number of parameters depends on the target architecture. This feature transformation merges "similar" gem5 parameters to get a statistical figure representing the entire processor which can have multiple cores. For example, merging the *number of branches* for each core creates a new feature describing the processor behavior.

6.5.2.4 Feature Combination

The feature transformation goal is to increase the data information gain by highlighting second-order correlations between features. By multiplying or adding arbitrary features, it is possible to emphasize any hidden relationships between variables. This function multiplies or adds two features (columns) from a dataframe using all possible column combinations. The generated columns are concatenated with the original DF.

6.5.3   Feature Selection

Selecting the relevant feature subset is crucial in any machine learning algorithms. Reducing the dataset improves the problem readability, shortens the training time, reduces the exploration dimensional space, and decreases the overfitting.

6.5.3.1   Variance Threshold

This technique removes features with low variability from the dataset (e.g., removing the features constant values). Further, the variance threshold method targets only the features (i.e., microarchitectural values) and not the labels (i.e., soft error analysis). If the feature has a lower variation, usually a constant value, it carries a reduced amount of information and can be removed.

6.5.3.2   Principal Component Analysis

Principal component analysis (PCA) is a mathematical method to reduce a dataset dimensionality by providing a set of orthogonal vectors indicating the maximum variance direction. Figure 6.2 shows random collections of points in orange and the black arrows represent the PCA components. The arrows indicate the direction and magnitude of the dataset variation when linearly dependent (Fig. 6.2a, b) while a random dataset provides an arbitrary direction (Fig. 6.2c). This work searches for the microarchitectural parameters with a significant impact on the application reliability. In this context, the PCA provides a fast approach to find linearly dependent variables that maximize a particular direction. By observing the components strength and direction (i.e., angle) it is possible to search the ones highlighting a strong correlation with the fault injection outcome. A greater X-axis variation (i.e., soft error) has more relevant information (Fig. 6.2a) than a group with higher correlation (Fig. 6.2b) with a high angle principal component (i.e., close to 90° or 0°).

6.5.3.3   Linear Regression

Linear regression models the relationship between two scalar variables, one dependent variable, and one explanatory variable. This technique searches for the best-fitting straight line through the data points usually using a least squares approach. By analyzing the model angle and accumulated error it is possible to rank the features (Y) with a stronger impact on the soft error assessment (X).

$$Y = \alpha X + \beta \tag{6.1}$$

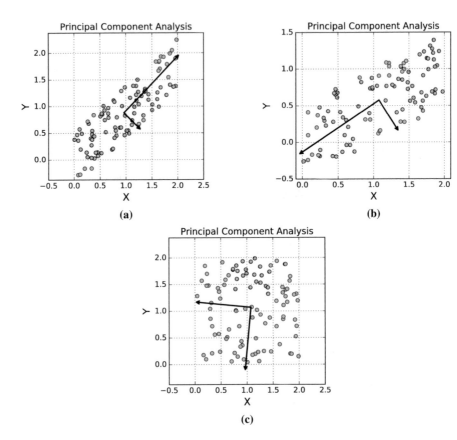

Fig. 6.2 Principal component analysis. (**a**) Strong linearly dependent. (**b**) Weak linearly dependent. (**c**) Random data

6.5.3.4   Correlation Coefficient

A correlation coefficient measures statistical dependence between two variables. In other words, this function describes two variables relationship through a score number between +1 and −1, where 0 represents the absence of correlation, 1 a complete correlation, and the sign gives the correlation direction. Pearson's is the most widely adopted correlation coefficient, and it measures linear relationships between two variables. Pearson's correlation coefficient is the covariance of the two variables divided by the product of their standard deviations.

$$\rho = \frac{\mathrm{cov}(X, Y)}{\sigma_x \sigma_y} \qquad \text{Pearson's correlation coefficient}$$

Further, it can be affected by strong outliers in a given dataset. For this purpose, Charles Spearman proposed a nonparametric measure of rank correlation to assess two variables monotonic relationship. A monotonic relationship means that two variables move towards the same relative direction, not necessary at an equal rate. By using ranked variables, the Spearman's coefficient presents a more robust solution to skewed points.

$$\rho_s = 1 - \frac{6 \sum d_i^2}{n(n^2 - 1)} \qquad \text{Spearman's correlation coefficient}$$

The coefficient selection is dependent on the raw dataset and its distribution. It is possible to filter features from a dataset by analyzing the ones with higher coefficients. Figure 6.3 shows four distinct datasets of 100 points considering the Spearman's and the Pearson's coefficient. First, Fig. 6.3a shows random points with no dependence. Both coefficients have a similar performance in linear dependence

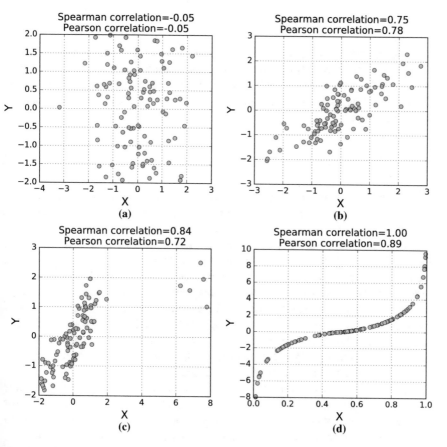

Fig. 6.3 Principal component analysis. (**a**) Random. (**b**) Linearly dependent. (**c**) Linearly dependent with outliers. (**d**) Non-linear relationship

scenarios (Fig. 6.3b), while the Pearson's suffers from strong outliers (Fig. 6.3c). Fig. 6.3d displays a more monotonic dataset, where Spearman's has a better score.

6.5.3.5   Recursive Feature Elimination

Recursive Feature Elimination (RFE) uses a given external estimator to prune a dataset until a target number of features. It employs the estimator coefficient to recursively remove features, creates a model, and calculates its accuracy. The Scikit-learn provides an RFE wrapper method for an arbitrary estimator. This work explored a linear regression, Huber Regressor, and among other estimators. The quality of the selection depends on the estimator capability to classify different features relevance for the accuracy model.

6.5.3.6   Euclidean Distance

The dataframe may contain (almost) identical values in the original data or arising from applied transformations. Removing such features improve the training time without reducing the dataset information. This method computes the Euclidean Distance between two features, removing it if they are closer than a predefined threshold.

6.5.3.7   Soft Error Score

Microarchitectural information (i.e., features) comes in multiple ways and they represent a wide range of parameters such as cache misses, # of float instructions, or virtual memory page size. Each one has a different distribution, and for example, the virtual memory page size has few possible constant values, while the # of float instructions ranges from zero to billions. Sometimes, parameters are constant for one application while fluctuates in another benchmark execution. Figure 6.4 shows six common feature arrangements found in the raw data from the gem5 microarchitectural statistics, for example, Fig. 6.4a displays a random value distribution. Figure 6.4b, c exhibits two linear relationships being the first one stronger, while Fig. 6.4d, e displays constant features (i.e., independent from the soft error). Finally, Fig. 6.4f demonstrates the average behavior of raw features, mixing some linear behavior, outliers, and constant values.

Applying one feature filtering algorithm does not provide the target results (i.e., the microarchitectural features with higher impact on the system soft error reliability). For instance, the correlation coefficients perform poorly on noise dataset (Fig. 6.4f) where the relationships exist among other types of data (e.g., constant values). The PCA can find the maximum growth direction in complex features. However, it can result in false positives depending on the data distribution (Fig. 6.4e). Regression model provides a more robust solution without presenting

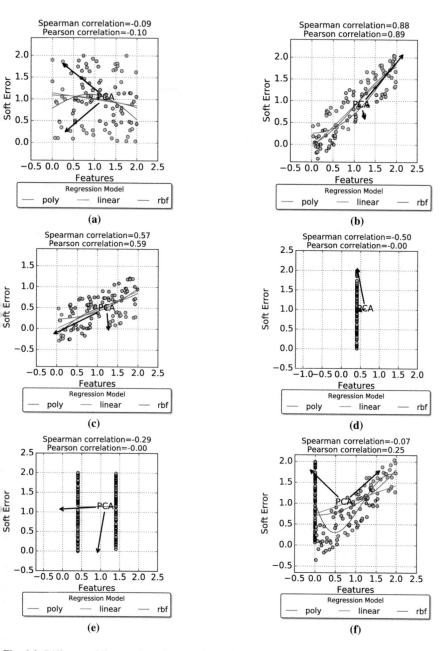

Fig. 6.4 Different of feature data shapes when using the gem5. (**a**) Random distribution. (**b**) Strong linear dependency. (**c**) Weak linear dependency. (**d**) Constant value. (**e**) Multiple constant values. (**f**) Usual feature shape

a general solution. Further, applying multiple filters in sequence results on a small dataset with a significant amount of false-positive solutions.

For this purpose, this work proposes a filter score (i.e., from 0 to 1) to measure the feature quality according to the target soft error problem. This score results from several algorithms simultaneous execution, in other words, the PCA, regression, and other techniques are computed in a single step. These outcomes are standardized in values between 0 and 1 to remove the parameter magnitude from the equation. The score computes the PCA first component, the Pearson's correlation coefficient, variance, dispersion score, and linear regression. Pearson's is preferable due to its smaller sensibility to false-positive linear relationships. Linear regression models with a 45° (from the origin) represent the optimal correlation between two dependent variables, increasing or decreasing this angle shows a weaker correlation.

The selection process computes the Soft Error Score for all columns in the target dataframe. The tool ranks the features by the score selecting the first N most significant values. This technique provides both filtering and ranking in a single technique, reducing the number of feature selection steps.

6.6   Exploration Flow

This book proposes an automated exploration flow to find the microarchitectural statistics that affect the most the fault injections outcome. Figure 6.5 shows this exploration execution flow which can be divided into three main phases.

6.6.1   Phase 1: Feature Acquisition and Data Homogenization

The tool searches for the input files and reads one by one, extracting the relevant information (Fig. 6.5a). Thus, two dataframes are created (b) where lines represent individual fault campaign scenarios (e.g., varying the # of cores, kernel, application) and the columns store the soft error (labels) or the microarchitectural statistics (features). The soft error DF is guaranteed to be complete (i.e., every cell has a value), however, the features depend on the scenario leaving unfilled data. Empty data points are replaced with NaN values (i.e., not a number) in the dataframe rectangular grid to retain a regular shape. Machine learning methods do not handle correctly abstract values, for this purpose, NaNs are replaced by zero in the features DF (c).

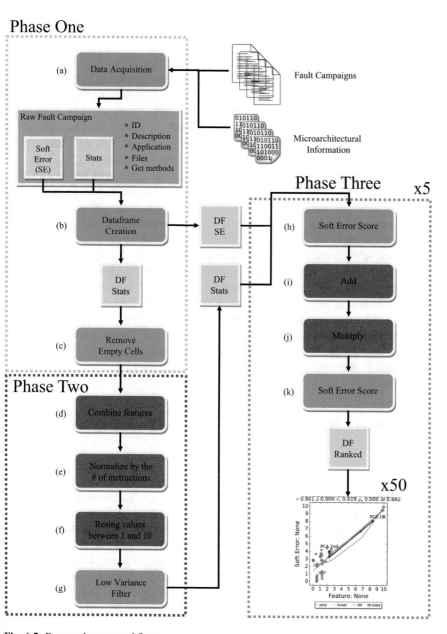

Fig. 6.5 Proposed automated flow

6.6.2   Phase 2: Unidimensional Feature Transformation and Selection

The second phase creates a new set of features by combining parameters from multiple cores. For instance, the gem5 report summarizes the # of float instructions by individual core (e.g., system.cpu0.num_fp_insts, system.cpu1.num_fp_insts, system.cpu2.num_fp_insts). By merging the multiple related features (Fig. 6.5d), it is possible to have the cores and the complete processor behavior. These reports provide most parameters as nominal values (e.g., system.cpu0.num_fp_insts, system.cpu0.op_class::MemWrite) which depends on the target application execution time (i.e., # of instructions). A direct comparison between two applications may not result in an evident relationship, as the distinct data magnitudes impact on the ML algorithm behavior. For instance, if an application A has 1000 branches and another B has 5000. Comparing the nominal value may lead to distinct conclusions, then if we analyze the branch participation (i.e., the number of branches divided by the total of executed instructions). To expand the exploration, the tool creates a second features DF where all columns are divided by the application # of instructions (e). Similarly, it is necessary to remove the magnitude variance from the features by resizing their range between 1 and 10 (f), enabling a fair comparison between features. This transformation also improves several machine learning techniques performance. Finally, the tool reduces the total number of features (i.e., independently from the soft error) by eliminating the ones with lower variance (g).

6.6.3   Phase 3: Multidimensional Feature Transformation and Selection

Until this point, the exploration flow was restricted to a single dimension (features). When considering the adopted soft error classification (i.e., vanish, hang, ONA, OMM, UT) the exploration becomes a five-dimensional problem. To reduce the computational cost, this flow performs five bi-dimensional investigations, each class against the features DF. First, the tool prunes the features from few thousand to 50 using the proposed Soft Error Score employing both features and labels (i.e., error classification) (Fig. 6.5h). Then, this reduced dataframe suffers two feature combination transformations: (i) addition and (j) multiplication. In other words, the tool adds and multiplies the fifty columns in every possible combination leading to a total of 5000 new features. Again, this significant dataset is pruned to the 50 most relevant features using the Soft Error Score which also provides the ranking of features (k). The tool automatically plots each feature and label with its Spearman's, Pearson's, and Soft Error Scores alongside the PCA and three regression models.

6.7 Results

6.7.1 Training Set Selection and Bias

Selecting a representative training set is paramount to have a meaningful result from any machine learning technique, i.e., the input dataset caries a bias leading to different outcomes. The algorithm accuracy depends on the acquired data, in other words, by judiciously reducing or increasing the training set it is possible to achieve different goals. This work collects inputs from multiple applications representing a large number of algorithms during a broad investigation. At the same time, the data has serial, OpenMP, and MPI benchmarks which can be broken down into distinct explorations with different results. Table 6.2 displays the available training sets.

To exemplify this behavior this work compares one parameter exploration by varying the training dataset. Additionally, this exploration targets the *Unexpected Termination* (UT) soft error class. Figure 6.6 shows the branch parameter using distinct inputs: All serial and parallel scenarios (Fig. 6.6a), OpenMP and MPI (Fig. 6.6b), only MPI (Fig. 6.6d), and only OpenMP (Fig. 6.6c). Considering every possible scenario, the # of branches has a not negligible relationship with the UT occurrence. However, both Spearman's and Pearson's describe a weak to medium correlation due to the prevalence of random data points. This interrelationship increases when restricting the ML inputs to parallel applications which are composed of MPI and OpenMP applications. Further, Fig. 6.6c, d breaks down the individual components of Fig. 6.6b, where the MPI applications visibly demonstrate a strong interaction between the number of branches and UTs. The three adopted coefficients (i.e., Spearman, Pearson, and soft error scores) exhibit this pattern by increasing 25% on average. By selection, the most appropriate training set is possible to reduce the error or focus on a smaller target population.

Table 6.2 Available training sets

Arch	Short	Description	Scenarios	Fault injections
V7	WCET	WCET serial	35	280,000
V7	MIB	Mibench serial	19	152,000
V7	SER	NASA NPB serial	10 10	160,000
V7	ROD	Rodinia OpenMP	16	128,000
V7	OMP	NASA NPB OpenMP	30 40	560,000
V7	MPI	NASA NPB MPI	25 32	456,000
V8	OMP	NASA NPB OpenMP	40	320,000
V8	MPI	NASA NPB MPI	32	256,000
Total			289	2,312,000

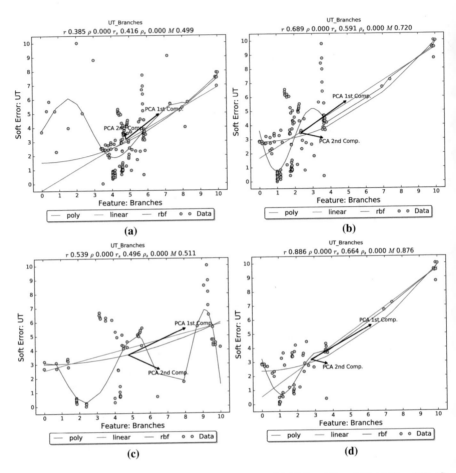

Fig. 6.6 Example of training set bias. (**a**) All scenarios. (**b**) MPI and OpenMP. (**c**) OpenMP. (**d**) MPI

6.7.2   Characterization

The following subsection discusses the characterization phase (Fig. 6.1c). We select the NASA NPB OpenMP and MPI benchmarks because they display a more interesting and realistic scenario considering multicore systems. The characterization precedes the machine learning execution, enabling the access to every parameter available. This phase enables the comparison between multiple applications in order to create subgroups. Also, it is possible to profile the target application in multiple execution scenarios (e.g., distinct compilation flags). Figure 6.7 shows these benchmarks soft error (red) and microarchitectural parameters (blue) divided by application (e.g., EP, LU). For this example, we selected six relevant parameters:

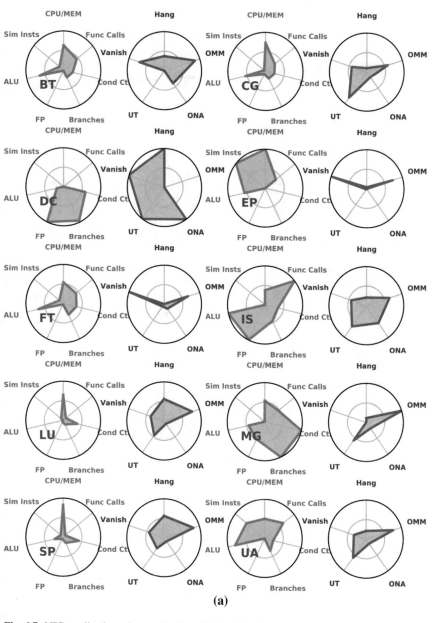

Fig. 6.7 NPB applications characterization. (**a**) OpenMP. (**b**) MPI

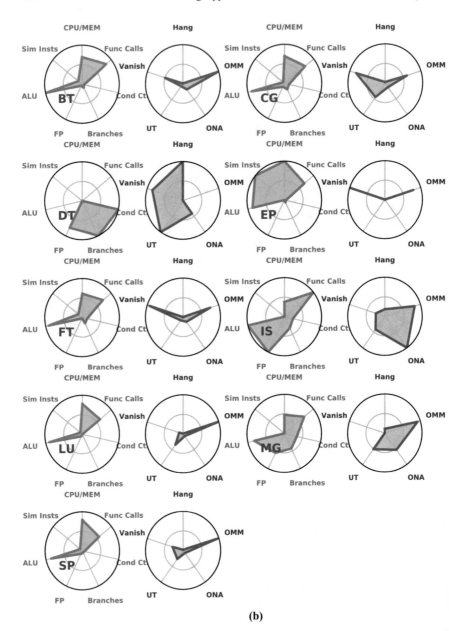

(b)

Fig. 6.7 (continued)

1. **CPU/MEM**: Its described the ratio between ALU operations by memory accesses (i.e., write and read). A larger number means a CPU-bound application and a smaller one a memory-bound.
2. **Func Calls**: Its described the concentration of function call instructions (i.e., the number of calls divided by the number of simulated instructions in total).
3. **Cond Ctrl**: The concentration of instructions with conditional control.
4. **Branches**: The concentration of branch instructions.
5. **FP**: The concentration of float-pointing instructions.
6. **Sim Insts**: Total number of simulated instructions.

This initial investigation phase provides some useful insights in training sets (e.g., OpenMP, MPI). It is visible that the large concentration of branch instructions in the OpenMP applications lead to a higher amount of UTs and hangs. This behavior is due to its parallel execution nature, as previously discussed. More significantly, it is possible to divide the application into subgroups among the datasets to explore similarities among the benchmarks. For instance, LU, MG, and SP are CPU-bounded with a high concentration of function calls. Further, the combination of both parameters correlates with the OMM (i.e., memory silent data corruptions) presence during fault campaigns. Next subsections explore the results extracted using the proposed tool also considering this characterization phase possible insights.

6.7.3   Branches and Function Calls

Control flow statements (e.g., for, if, else, while) are the fundamental bedrock of any programming language. The ability of conditionally executing code portions enables software engineers to create more complex algorithms where jump, branch, and function call instructions fulfill the control flow function at the assembly level. Any fault in such instructions typically causes unrecoverable errors (e.g., segmentation faults). The tool provides preliminary results for investigating the impact of control flow instruction in the application reliability. Figure 6.8 shows the effect of the branch instructions concentration (i.e., the number of branches divided by the total of instructions) on the soft error vulnerability considering multiple training sets. None of the fault injection classes demonstrates a strong correlation with the branch concentration due to the diversity of input scenarios such as serial and parallel applications, with execution time varying from 10 million to 87 billion instructions, from one to eight cores.

The training sets more comprehensive range conceals several individual relationships, in particular, how multicore applications behave under these conditions. For this purpose, Fig. 6.9 displays the branch instruction concentration focusing on the NPB MPI (Fig. 6.9b) and OpenMP-based (Fig. 6.9a) applications which presents a more extensive workload and better scalability. The execution time of the NAS suite ranges from 300 million to 87 billion instructions, from one to eight cores. Note that

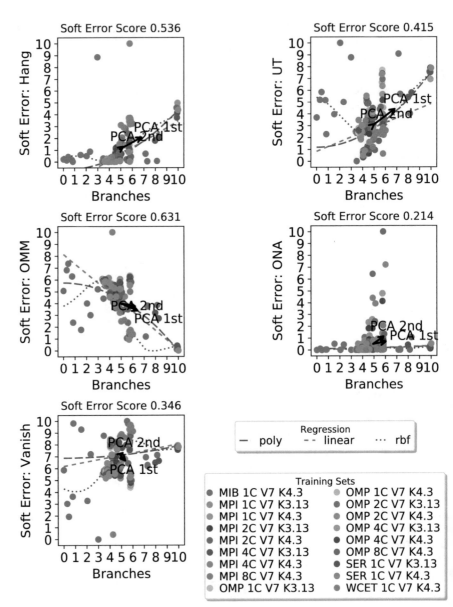

Fig. 6.8 Branch instruction impact on the soft error classification

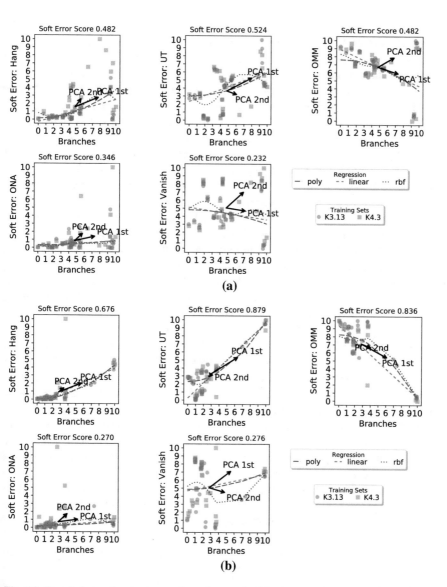

Fig. 6.9 Branch instruction impact on the soft error classification targeting parallel applications. (a) NPB OpenMP. (b) NPB MPI

the host Linux kernel (i.e., 4.3 or 3.13) shows no direct impact on the application reliability for this particular set of applications. Due to the NPB longer execution time (i.e., 16 billion instructions on average) only a fraction of this total is dedicated to the kernel. Consequently, direct faults injections in the Linux inner workings are extremely rare.

Considering the impact of the number of branches on the software stack reliability (Fig. 6.9), note how the OpenMP benchmarks exhibit a weaker relationship than the MPI ones. OpenMP and MPI have distinct behaviors under fault injection, for example, the OpenMP relies on loop (i.e., for-while) parallelizations leading to a greater branch presence. In contrast, MPI applications show a more considerable influence of the number of branches on its soft error vulnerability, especially when considering the Hang, UT, and OMM classes. Hang errors arise due to severe control flow errors (e.g., incorrect iteration counter), UT is an unexpected application termination (e.g., wrong address calculation), and an OMM results from an application finishing with an incorrect memory. Both Hang and UT show a direct and positive correlation while OMM a negative correlation, in other words, the occurrence of Hang and UT increases at the same time that the OMM reduces considering MPI applications. In this case, branch errors lead to Hangs or UTs which in other circumstance would finish with an erroneous memory.

Function calls are a particular type of control flow instruction which simultaneously changes the program counter and stack pointer. Observing the Hang error (see Fig. 6.10), both MPI and OpenMP reliability improves when increasing the number of function call until a saturation level. The UT class follows the Hang behavior (Fig. 6.10b) while the OMM presents an inverse correlation. Additionally, the distinct effect of the OMM versus Hang and UT is similar to the one displayed by the branch on MPI-based benchmarks (Fig. 6.9b). Analyzing individual parameters not always expose direct relationships between profiling data and fault injection campaigns. By combining both figures, it is possible to uncover a strong correlation between this new index value (i.e., number of function calls times the number of branches) and the error classes. Figure 6.11 exemplifies this behavior, note that this new index value and the Hang percentage increases simultaneously, an observable behavior through several scenarios for both MPI and OpenMP. The tool main ability is to combine multiple features and highlight previously unknown correlations leading to a profound understatement of the scenarios behavior.

As previously mentioned, particular applications can be grouped into subgroups according to their similarity. LU, MG, and SP form one of such collections as they are CPU-bounded with a high concentration of function calls. The three applications reliability (i.e., vanish rate) improves as the index (i.e., function calls times the branches) grows, also noting how the OMM occurrence also decreases in Fig. 6.12a. Reads and write operations in memory-bounded algorithms usually move these fault values to the memory. Instead, their CPU-bounded nature leads to longer vulnerability windows, i.e., the data remains in the exposed register file longer. In this particular applications, function calls have a rejuvenating effect by restoring old registers values saved during earlier iterations. This behavior is more explicit in Fig. 6.12b by visualizing the same data without resizing the soft error (Y) axis.

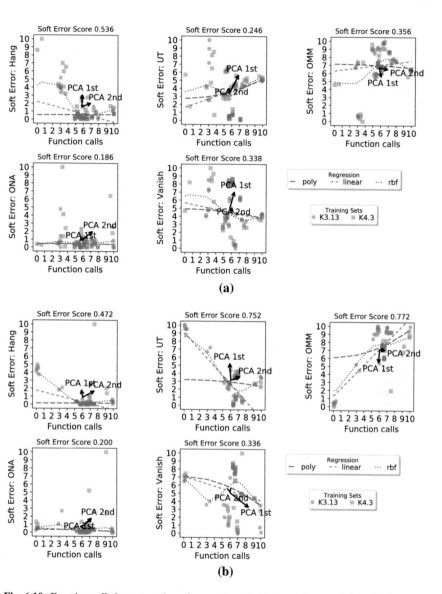

Fig. 6.10 Function calls impact on the soft error classification targeting parallel applications. (**a**) NPB OpenMP. (**b**) NPB MPI

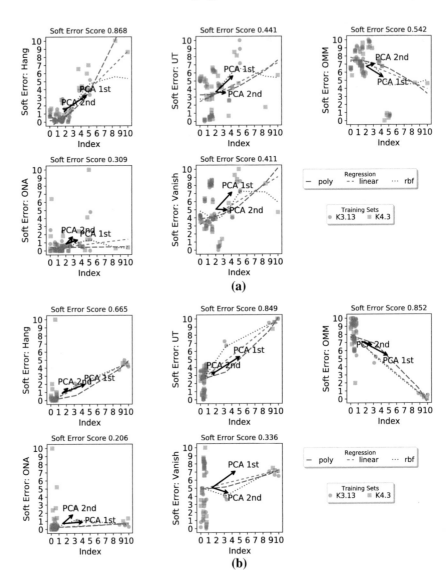

Fig. 6.11 Proposed index impact on the soft error classification targeting parallel applications. (**a**) NPB OpenMP. (**b**) NPB MPI

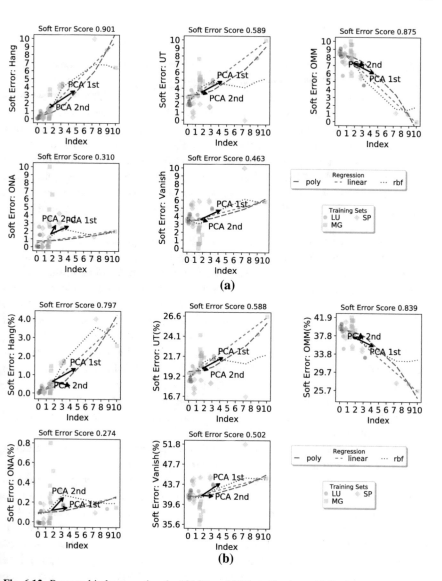

Fig. 6.12 Proposed index targeting the LU,SP, and MG applications. (**a**) Resizing both axis. (**b**) Unmodified soft error values

6.7.4 Memory Transactions

At instruction level on a RISC processor, the address generation of memory access operations (e.g., load and stores) can be compromised by soft errors as the source register faults lead to wrong address calculations. The reduced number of ARMv7 registers to perform address calculations leads to the use of load–store templates by the compiler to diminish the computational cost of register recycling. In other words, the ARMv7 compiler continuously utilizes the same register to perform memory transactions (e.g., R0–3 and SP). Figure 6.13 shows the soft error results (e.g., Vanish+OMM+ONA, UT, Hangs) alongside the memory access figures both integer and float-pointing (FP). Vanish, OMM, and ONA were clustered into a single group to investigate the UT and Hang incidence. Increasing the number of load–store operations can lead to a more significant UT and Hang occurrence in MPI applications using an OS on top of the ARMv7 processor (Fig. 6.13b). In applications such as MG and IS the increasing percentage of memory transactions (i.e., load and stores instructions) display a growing occurrence of hang and UT. For example, MG MPI application memory-oriented operations for single and quad-core processors are 15 and 22%, while the UT occurrence increases from 22 to 30%. DT and DC have a less pronounced behavior due to its limited execution time approximately 500 million instructions while the NPB average floats around 16 billion. In contrast, the OpenMP-based applications exhibit no direct impact of the # of load–stores in the Hang/UT incidence. The OpenMP library parallelizes the *for* statements and to achieve a higher performance the loop index remains on the register file (i.e., reducing the memory access for branch executions). In Fig. 6.13a varying the number of memory accesses has no apparent correlation with the UT and Hang occurrence on OpenMP benchmarks.

The ARMv7 workload for a single faultless execution has an instruction count that ranges from 299 million to 87 billion, with an average of 16 billion of instructions. In contrast, the 64-bit architecture applications execute in average 654 million instructions, varying from 41 million to 3 billion. Applications executed using the ARMv8 ISA present a significant performance improvement when compared to the ARMv7. This performance gain can be pinpointed to the removal of several legacy features (e.g., fast and multilevel interrupts, conditional instructions) and to significant improvements in the floating-point unit by adding new specialized instructions and increasing the FP register file. The ARMv7 often resorts to the ARM software FP library to perform some operations and thus increasing execution time due to automatically compiler decisions. The 64-bit architecture exhibits a similar behavior considering FP memory transactions, supporting the claim above that wrong address calculation related to memory access, as FP instructions are exclusively used for computation and not for control flow operations (e.g., branches and jumps). The ARMv8 architecture has a higher percentage of float-pointing instructions than the ARMv7. Figure 6.14 displays several scenarios of soft error analysis and FP memory figures. Reducing the memory transactions participation from the total number of executed instructions for LU and SP applications show

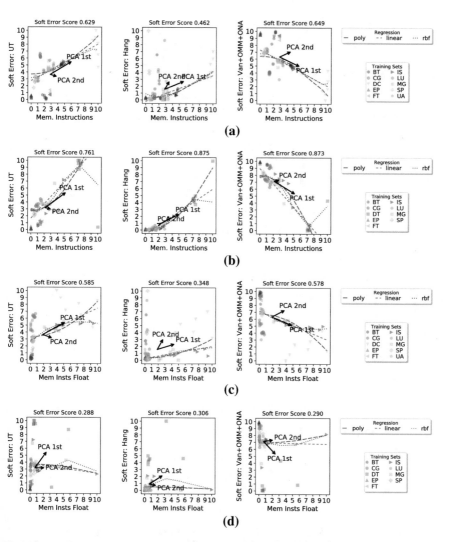

Fig. 6.13 Integer and Float-pointing (FP) memory access instructions impact on the soft error classification targeting parallel applications. (**a**) NPB OpenMP integer mem. access. (**b**) NPB MPI integer mem. access. (**c**) NPB OpenMP FP mem. access. (**d**) NPB MPI FP mem. access

a UT occurrence reduction trend. FT and UA MPI applications reinforce this hypothesis by demonstrating that a constant memory-oriented instruction incidence leads to a regular UT percentage.

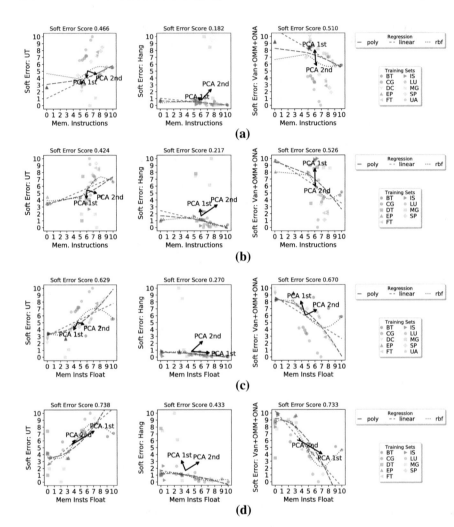

Fig. 6.14 Integer and Float-pointing (FP) memory access instructions impact on the soft error classification using a 64-bit ARMv8 architecture. (**a**) NPB OpenMP integer mem. access. (**b**) NPB MPI integer mem. access. NPB OpenMP FP mem. access. (**d**) NPB MPI FP mem. access

6.8   Case Study

Following Fig. 6.1, this case study first selects and designs the target application. Nowadays, self-driven cars capable of automatically steering the vehicle are being tested. Such systems should be able to analyze the real word, make decisions, and perform actions resulting in a complex software stack with multiple algorithms. Visual odometry (VO) is one of the most critical subsystems, which is the process of determining the position and orientation of a robot by analyzing the associated

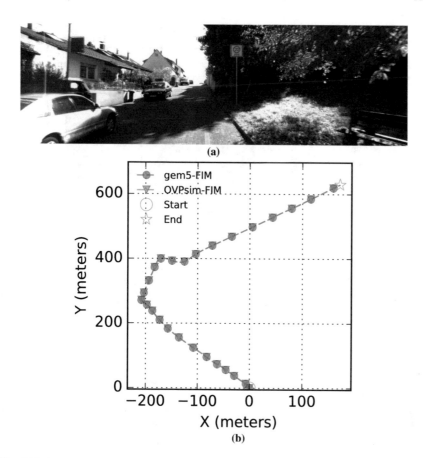

Fig. 6.15 Selected visual odometry (VO) application extracted from the KITTI Vision Benchmark Suite [40, 41]. (**a**) Input frame from KITTI benchmark 11. (**b**) VO application output path using distinct virtual platforms

camera images. In other words, the VO system uses a sequence of images to determine the vehicle traveled path. This work applies the acquired knowledge with multiple characterized benchmarks and the proposed machine learning tool in a real-life scenario. For this case study, we selected the LIBVISO2 [41] visual odometry library developed by the Karlsruhe Institute of Technology (KIT). The library and dependencies are compiled using the arm GCC cross-compiler with hard float-pointing and single instruction, multiple data (SIMD) flags enabled. Our LIBVISO2 setup uses the KITTI Vision Benchmark Suite [40] which is composed of 22 stereo odometry sequences from real-life vehicle trajectories. Figure 6.15 shows two input frames of the benchmark 11 from the KITTI suite alongside the algorithm predicted path and ground truth.

The initial exploration comprises four fault injection campaigns targeting the benchmark 11, each one compiles with a different optimization flag. Further, this first investigation is restricted to 100 frames from a total of 920 where the

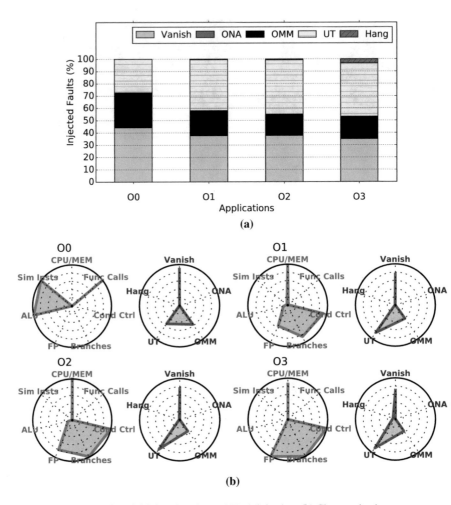

Fig. 6.16 Benchmark 11 initial exploration. (**a**) Fault injection. (**b**) Characterization

fault campaigns are performed by the OVPsim-FIM as described in Sect. 3.5. The OVPsim-FIM presents a flexible and fast fault injection framework enabling more complex soft error vulnerability explorations. While the characterization phase deploys the gem5-FIM (Sect. 3.5) to extract the microarchitectural parameters, which is also necessary for the machine learning investigation. The chosen VPs have differences in the simulation engine without affecting the target algorithm produced output. Figure 6.15b shows the benchmark 11 execution using the OVPsim-FIM and gem5-FIM which are identical. Figure 6.16 shows this first exploration fault injection (a) and characterization (b) results. Note how the application without any optimization (i.e., O0) has an execution 4.7x times larger than any optimization flag (i.e., O1–3), from 150 to 32 billion instructions, respectively. This extended execution time results on a more significant number of register operations (ALU) as shown by Fig. 6.16b. Because the compiler optimization reduces the number of instructions between control flow statements (e.g., if) it increases the probability of Hangs and Unexpected Termination. When comparing the application with code optimization (i.e., O1–3), it is noticeable a growing number of hang occurrences. For example, observe the growing number of branches with more aggressive GCC optimizations.

The visual odometry benchmark was chosen due to its real-life applicability in self-driven cars which enables a physical representation of the system reliability (i.e., the traveled path). Figure 6.17 depicts the traveled way (A), # of evaluated frames (B), and the deviation between the correct and predicted stop point (C) for each algorithm simulation under fault injection. Lastly, Fig. 6.17 (D) shows the scenario (C) restricted to completed executions (i.e., all frames were evaluated by the algorithm). Figure 6.17a exhibits these four plots considering no GCC optimization (O0). In this scenario, for instance, 73.12% of the fault injections terminate (i.e., evaluates 100% of the frames), from whom only four are incorrect. In contrast, when using compiler optimizations, the number of completed simulations decreases from 73.12 to 55%. In this example (Fig. 6.17) the vehicle travels around 70 m after processing 100 frames. The error reaches up to 70, where larger values are due to algorithm halts (i.e., UT or Hang) stopping far from the correct point. The error, considering only completed executions, deviates up to 0.50 meters when compiled with optimizations while the O0 flag has no substantial deviation from the correct path.

Altering the GCC optimization flags during the compilation is one possible solution to change the application characteristics. Each flag should change the compiler algorithm to improve assembly code in a specific manner. For example, the unroll loops option reduces the number of loop iterations, consequently, reducing the number of branch instructions. The next experiments explore the capabilities of the GCC to influence the application characteristics aiming to improve its reliability. Compilers, such as the GCC, have optimization flags enabled by default even when using the argument "-O0." The first fault campaign as depicted in Fig. 6.19a shows the benchmark 11 compiled without any optimization flag under the influence of fault injections. Between the O0 flag and no optimization flag, the final object code has not meaningful changes leading to almost identical simulations (i.e., both application execute 150 billion instructions). Figures 6.17a and 6.19a display this similarity considering the four parameters A–D.

Previous results show a considerable reliability degradation when compiling with any other compilation (i.e., O1–3) while the O0 has a notable speed reduction (i.e., 4.7×). Further, the most noticeable changes occur when transitioning from O0 to O1, and thus, we investigate how to improve the O0 performance by adding new compilation flags. The target application heavenly depends on loops to analyze the images, for this purpose, some unrolled loops options have been added to its compilation (Fig. 6.19b). The next experiment (Fig. 6.19c) mimics the O3 option by using individual optimization flags instead (e.g., -falign-jumps -falign-loops -falign-labels -fcaller-saves -fcrossjumping). This investigation aims to replicate the O3 effect on the code without directly adding the "-O3" flag. Individual optimization flags have minimal impact on the final system reliability/performance as they have a minor impact on the compiler behavior. The main optimization flags (e.g., O0-3)

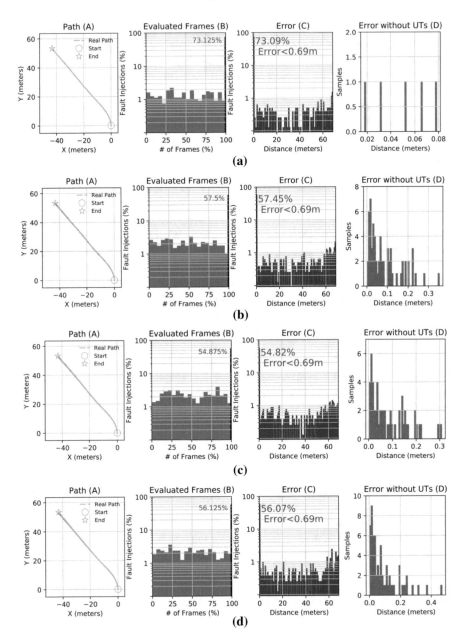

Fig. 6.17 Benchmark 11 initial exploration traveled route. (**a**) O0. (**b**) O1. (**c**) O2. (**d**) O3

are not just group alias, and they have a direct impact on the compiler algorithm. In other words, the GCC does not provide full control of these optimizations leaving the most relevant options hardcoded in the compiler source code. For example, the

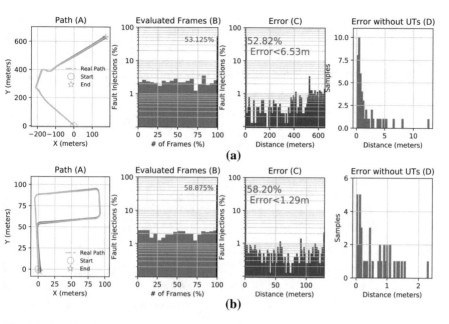

Fig. 6.18 Benchmarks complete execution. (**a**) Benchmark 11. (**b**) Benchmark 14

benchmark 11 without any optimization simulates 150 billion instructions while the O3 recreation reduces this amount by only 30 billion when considering 100 frames, still four times longer than a similar scenario compiled with the O1 flag.

After analyzing the impact of multiple GCC flags on the visual odometry application, we extended this exploration to the effects of the traveled path on the accumulated error. This exploration uses two input sets from the KITTI suite, the 11 featuring 920 frames and the 14 with 630 frames, and both compiled with the O3 optimization flag. Figure 6.18 shows the traveled path for each fault injection scenario (A) with a zoom on the final stopping point (B). The benchmark 11 follows a more straight path leading to horizontal errors, in other words, the vehicle diverges either to the right or the left (Fig. 6.18a). In benchmark 14, each traveled curve corrects the application towards the correct path (dashed green line) as right and left accumulated errors counteract each other. However, the vertical error accumulates in a greater magnitude as shown by Fig. 6.18b. Figure 6.18 also displays the distance between the car stopping the correct one; considering every fault injection (Fig. 6.18c) and only the ones with complete (i.e., neither Hang nor UT) application (Fig. 6.18d). Note how the number of completed simulations (in red) remains at the same levels of the reduced simulations (i.e., 100 frames), around 55% when compiled with an identical GCC flag (i.e., O3).

Most of the existing software projects use O2 as standard shipping optimization flag because a four times slowdown is not acceptable. In the impossibility of improving the software compiled with O0, the last experiment attempts to reduce the number of unexpected terminations when using the O3 option in the GCC. This fault campaign (Fig. 6.19d) uses aggressive loop and branch optimizations,

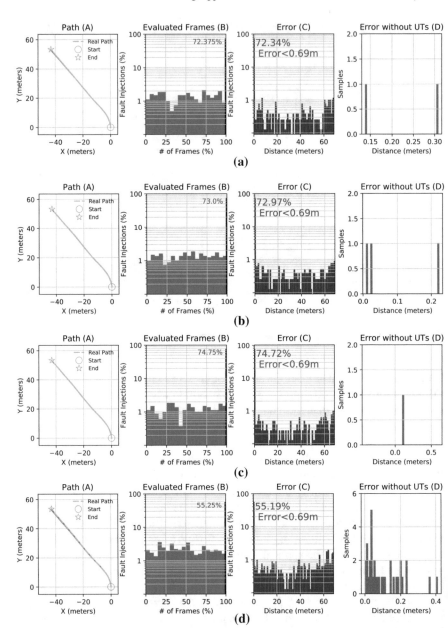

Fig. 6.19 Exploring the gcc optimization flags. (**a**) No optimization (i.e., removing all possible optimizations flags including the default ones). (**b**) O0 plus multiple unrolling loop options. (**c**) Fabricated O3. (**d**) O3 plus multiple unrolling loop options

given more freedom to the compiler to allocate the instructions orders. Again, it demonstrates the GCC lack of support to algorithm customizations as most of the code transformations are locked to hardcoded arguments.

6.9   Closing Remarks

This chapter introduced the basic concepts of machine learning and its applicability to system development. Further, we reviewed the state-of-the-art ML approaches to improve the reliability assessment of large-scale systems. By observing the lack of methodologies targeting embedded multicore systems software stack, this work proposes the use of ML algorithms to speed up dependability investigations during early design space explorations. We proposed a cross-layer investigation tool which performs multivariable and statistical analyses using the gem5 microarchitectural information combined with fault injection campaigns. This tool enables reducing the number of fault injection campaigns or improving the mitigation technique efficiency by observing microarchitectural symptoms. To explore the functionalities of the promoted development cycle, we investigated the reliability of a visual odometry algorithm common on self-driven cars.

References

1. R. Akerkar, P.S. Sajja, *Intelligent Techniques for Data Science* (Springer, Berlin, 2016). https://www.springer.com/gp/book/9783319292052
2. M.A. Alam, H. Kufluoglu, D. Varghese, S. Mahapatra, A comprehensive model for PMOS NBTI degradation: recent progress. Microelectron. Reliab. **47**(6), 853–862 (2007). https://doi.org/10.1016/j.microrel.2006.10.012
3. M. Alian, D. Kim, N.S. Kim, pd-gem5: simulation infrastructure for parallel/distributed computer systems. IEEE Comput. Archit. Lett. **15**(1), 41–44 (2016). https://doi.org/10.1109/LCA.2015.2438295
4. ARM: Technologies | big.LITTLE (2017). https://developer.arm.com/technologies/big-little
5. L. Artola, G. Hubert, R. Schrimpf, Modeling of radiation-induced single event transients in SOI FinFETS, in *2013 IEEE International Reliability Physics Symposium (IRPS)* (2013), pp. SE.1.1–SE.1.6. https://doi.org/10.1109/IRPS.2013.6532108
6. R.A. Ashraf, R. Gioiosa, G. Kestor, R.F. DeMara, C.Y. Cher, P. Bose, Understanding the propagation of transient errors in HPC applications, in *Proceedings of the International Conference for High Performance Computing, Networking, Storage and Analysis, SC '15* (ACM, New York, 2015), pp. 72:1–72:12. https://doi.org/10.1145/2807591.2807670
7. T. Austin, E. Larson, D. Ernst, SimpleScalar: an infrastructure for computer system modeling. Computer **35**(2), 59–67 (2002). https://doi.org/10.1109/2.982917
8. D.H. Bailey, E. Barszcz, J.T. Barton, D.S. Browning, R.L. Carter, L. Dagum, R.A. Fatoohi, P.O. Frederickson, T.A. Lasinski, R.S. Schreiber, H.D. Simon, V. Venkatakrishnan, S.K. Weeratunga, The NAS parallel benchmarks summary and preliminary results, in *Proceedings of the 1991 ACM/IEEE Conference on Supercomputing (Supercomputing '91)* (1991), pp. 158–165 https://doi.org/10.1145/125826.125925
9. J. Baraza, J. Gracia, D. Gil, P. Gil, A prototype of a VHDL-based fault injection tool, in *Proceedings of the IEEE International Symposium on Defect and Fault Tolerance in VLSI Systems, 2000* (2000), pp. 396–404. https://doi.org/10.1109/DFTVS.2000.887180
10. R. Baumann, Radiation-induced soft errors in advanced semiconductor technologies. IEEE Trans. Device Mater. Reliab. **5**(3), 305–316 (2005). https://doi.org/10.1109/TDMR.2005.853449
11. L. Bautista-Gomez, F. Zyulkyarov, O. Unsal, S. McIntosh-Smith, Unprotected computing: a large-scale study of DRAM raw error rate on a supercomputer, in *International Conference for High Performance Computing, Networking, Storage and Analysis, SC16* (2016), pp. 645–655. https://doi.org/10.1109/SC.2016.54

12. F. Bellard, QEMU, A fast and portable dynamic translator, in *Proceedings of the Annual Conference on USENIX Annual Technical Conference* (2005), pp. 41–41. http://dl.acm.org/citation.cfm?id=1247360.1247401

13. G. Beltrame, C. Bolchini, L. Fossati, A. Miele, D. Sciuto, A framework for reliability assessment and enhancement in multi-processor systems-on-chip, in *22nd IEEE International Symposium on Defect and Fault-Tolerance in VLSI Systems (DFT 2007)* (2007), pp. 132–142. https://doi.org/10.1109/DFT.2007.35

14. G. Beltrame, C. Bolchini, A. Miele, Multi-level fault modeling for transaction-level specifications, in *Proceedings of the 19th ACM Great Lakes Symposium on VLSI, GLSVLSI '09* (ACM, New York, 2009), pp. 87–92. https://doi.org/10.1145/1531542.1531565

15. G. Beltrame, L. Fossati, ReSP: a design and validation tool for data systems, in *Data Systems in Aerospace, DASIA 2008*, vol. 665 (2008), p. 22. http://adsabs.harvard.edu/abs/2008ESASP.665E..22B

16. C. Bienia, S. Kumar, J.P. Singh, K. Li, The PARSEC benchmark suite: characterization and architectural implications, in *Proceedings of the 17th International Conference on Parallel Architectures and Compilation Techniques, PACT '08* (ACM, New York, 2008), pp. 72–81. https://doi.org/10.1145/1454115.1454128

17. D. Binder, E.C. Smith, A.B. Holman, Satellite anomalies from galactic cosmic rays. IEEE Trans. Nucl. Sci. **22**(6), 2675–2680 (1975). https://doi.org/10.1109/TNS.1975.4328188

18. N. Binkert, B. Beckmann, G. Black, S.K. Reinhardt, A. Saidi, A. Basu, J. Hestness, D.R. Hower, T. Krishna, S. Sardashti, R. Sen, K. Sewell, M. Shoaib, N. Vaish, M.D. Hill, D.A. Wood, The Gem5 simulator. SIGARCH Comput. Archit. News **39**(2), 1–7 (2011). https://doi.org/10.1145/2024716.2024718

19. R. Bishop, *Intelligent Vehicle Technology and Trends* (Artech House, Boston, 2005)

20. S. Borkar, A.A. Chien, The future of microprocessors. Commun. ACM **54**(5), 67 (2011). https://doi.org/10.1145/1941487.1941507. http://cacm.acm.org/magazines/2011/5/107702-the-future-of-microprocessors/fulltext

21. S. Borkar, T. Karnik, S. Narendra, J. Tschanz, A. Keshavarzi, V. De, Parameter variations and impact on circuits and microarchitecture, in *Proceedings 2003. Design Automation Conference (IEEE Cat. No.03CH37451)* (2003), pp. 338–342. https://doi.org/10.1145/775832.775920

22. D. Borodin, B.H. Juurlink, Protective redundancy overhead reduction using instruction vulnerability factor, In: *Proceedings of the Seventh ACM International Conference on Computing Frontiers, CF '10* (ACM, New York, 2010), pp. 319–326. https://doi.org/10.1145/1787275.1787342

23. S. Buchanan, NOAA kicks off 2018 with massive supercomputer upgrade. National Oceanic and Atmospheric Administration (2018). http://www.noaa.gov/media-release/noaa-kicks-off-2018-with-massive-supercomputer-upgrade

24. A. Butko, R. Garibotti, L. Ost, G. Sassatelli, Accuracy evaluation of GEM5 simulator system, in *2012 Seventh International Workshop on Reconfigurable Communication-Centric Systems-on-Chip (ReCoSoC)* (2012), pp. 1–7. https://doi.org/10.1109/ReCoSoC.2012.6322869

25. S. Che, M. Boyer, J. Meng, D. Tarjan, J.W. Sheaffer, S.H. Lee, K. Skadron, Rodinia: a benchmark suite for heterogeneous computing, in *2009 IEEE International Symposium on Workload Characterization (IISWC)* (2009), pp. 44–54. https://doi.org/10.1109/IISWC.2009.5306797

26. H. Cho, S. Mirkhani, C.Y. Cher, J. Abraham, S. Mitra, Quantitative evaluation of soft error injection techniques for robust system design, in *2013 50th ACM/EDAC/IEEE Design Automation Conference (DAC)* (2013), pp. 1–10

27. M. Didehban, A. Shrivastava, nZDC: a compiler technique for near zero silent data corruption, in *Proceedings of the 53rd Annual Design Automation Conference, DAC '16* (ACM, New York, 2016), pp. 48:1–48:6. https://doi.org/10.1145/2897937.2898054

28. F. de Aguiar Geissler, F. Lima Kastensmidt, J. Pereira Souza, Soft error injection methodology based on QEMU software platform, in *Test Workshop—LATW, 2014 15th Latin American* (2014), pp. 1–5. https://doi.org/10.1109/LATW.2014.6841910
29. B. de Dinechin, R. Ayrignac, P.E. Beaucamps, P. Couvert, B. Ganne, P. de Massas, F. Jacquet, S. Jones, N. Chaisemartin, F. Riss, T. Strudel, A clustered manycore processor architecture for embedded and accelerated applications, in *2013 IEEE High Performance Extreme Computing Conference (HPEC)* (2013), pp. 1–6. https://doi.org/10.1109/HPEC.2013.6670342
30. D. DiTomaso, T. Boraten, A. Kodi, A. Louri, Dynamic error mitigation in NoCs using intelligent prediction techniques, in *2016 49th Annual IEEE/ACM International Symposium on Microarchitecture (MICRO)* (2016), pp. 1–12. https://doi.org/10.1109/MICRO.2016.7783734
31. A. Dixit, A. Wood, The impact of new technology on soft error rates, in *2011 International Reliability Physics Symposium (2011)*, pp. 5B.4.1–5B.4.7. https://doi.org/10.1109/IRPS.2011.5784522
32. P.E. Dodd, F.W. Sexton, Critical charge concepts for CMOS SRAMs. IEEE Trans. Nucl. Sci. **42**(6), 1764–1771 (1995). https://doi.org/10.1109/23.488777
33. B. Doyle, B. Boyanov, S. Datta, M. Doczy, S. Hareland, B. Jin, J. Kavalieros, T. Linton, R. Rios, R. Chau, Tri-gate fully-depleted CMOS transistors: fabrication, design and layout, in *2003 Symposium on VLSI Technology, 2003*. Digest of Technical Papers (2003), pp. 133–134. https://doi.org/10.1109/VLSIT.2003.1221121
34. A. Ebnenasir, R. Hajisheykhi, S.S. Kulkarni, Facilitating the design of fault tolerance in transaction level systemC programs, in *Distributed Computing and Networking*, ed. by L. Bononi, A.K. Datta, S. Devismes, A. Misra, vol. 7129. Lecture Notes in Computer Science (Springer, Berlin, 2012), pp. 91–105
35. M. Ebrahimi, H. Asadi, R. Bishnoi, M.B. Tahoori, Layout-based modeling and mitigation of multiple event transients. IEEE Trans. Comput. Aided Des. Integr. Circuits Syst. **35**(3), 367–379 (2016). https://doi.org/10.1109/TCAD.2015.2459053
36. H. Esmaeilzadeh, E. Blem, R. St. Amant, K. Sankaralingam, D. Burger, Power limitations and dark silicon challenge the future of multicore. ACM Trans. Comput. Syst. **30**(3), 11:1–11:27 (2012). https://doi.org/10.1145/2324876.2324879.
37. H. Esmaeilzadeh, E. Blem, R. St. Amant, K. Sankaralingam, D. Burger, Dark silicon and the end of multicore scaling, in *2011 38th Annual International Symposium on Computer Architecture (ISCA)* (2011), pp. 91–105
38. R. Garibotti, L. Ost, R. Busseuil, M. Kourouma, C. Adeniyi-Jones, G. Sassatelli, M. Robert, Simultaneous multithreading support in embedded distributed memory MPSoCs, in *2013 50th ACM/EDAC/IEEE Design Automation Conference (DAC)* (2013), pp. 1–7. https://doi.org/10.1145/2463209.2488836
39. A. Geiger, P. Lenz, R. Urtasun, Are we ready for autonomous driving? The KITTI vision benchmark suite, in *2012 IEEE Conference on Computer Vision and Pattern Recognition* (2012), pp. 3354–3361. https://doi.org/10.1109/CVPR.2012.6248074
40. A. Geiger, P. Lenz, R. Urtasun, Are we ready for autonomous driving? The Kitti vision benchmark suite, in *Conference on Computer Vision and Pattern Recognition (CVPR)* (2012)
41. A. Geiger, J. Ziegler, C. Stiller, Stereoscan: dense 3D reconstruction in real-time, in *Intelligent Vehicles Symposium (IV)* (2011)
42. I. Giurgiu, J. Szabo, D. Wiesmann, J. Bird, Predicting DRAM reliability in the field with machine learning, in *Proceedings of the 18th ACM/IFIP/USENIX Middleware Conference: Industrial Track, Middleware '17* (ACM, New York, 2017), pp. 15–21. https://doi.org/10.1145/3154448.3154451
43. T. Granlund, B. Granbom, N. Olsson, Soft error rate increase for new generations of SRAMs. IEEE Trans. Nucl. Sci. **50**(6), 2065–2068 (2003). https://doi.org/10.1109/TNS.2003.821593
44. W. Gropp, R. Thakur, E. Lusk, *Using MPI-2: Advanced Features of the Message Passing Interface*, 2nd edn. (MIT Press, Cambridge, 1999)

45. Q. Guan, N. BeBardeleben, P. Wu, S. Eidenbenz, S. Blanchard, L. Monroe, E. Baseman, L. Tan, Design, Use and evaluation of P-FSEFI: a parallel soft error fault injection framework for emulating soft errors in parallel applications, in *Proceedings of the 9th EAI International Conference on Simulation Tools and Techniques, SIMUTOOLS'16* (Institute for Computer Sciences, Social-Informatics and Telecommunications Engineering (ICST), Brussels, 2016), pp. 9–17. http://dl.acm.org/citation.cfm?id=3021426.3021429

46. Q. Guan, N. Debardeleben, S. Blanchard, S. Fu, F-SEFI: a fine-grained soft error fault injection tool for profiling application vulnerability, in *2014 IEEE 28th International Parallel and Distributed Processing Symposium* (2014), pp. 1245–1254. https://doi.org/10.1109/IPDPS.2014.128

47. M. Guthaus, J. Ringenberg, D. Ernst, T. Austin, T. Mudge, R. Brown, MiBench: a free, commercially representative embedded benchmark suite, in *2001 IEEE International Workshop on Workload Characterization, WWC-4* (2001), pp. 3–14. https://doi.org/10.1109/WWC.2001.990739

48. A. Gutierrez, J. Pusdesris, R.G. Dreslinski, T. Mudge, C. Sudanthi, C.D. Emmons, M. Hayenga, N. Paver, Sources of error in full-system simulation, in *2014 IEEE International Symposium on Performance Analysis of Systems and Software (ISPASS)* (2014), pp. 13–22. https://doi.org/10.1109/ISPASS.2014.6844457

49. S.K.S. Hari,, S.V. Adve, H. Naeimi, P. Ramachandran, Relyzer: Exploiting application-level fault equivalence to analyze application resiliency to transient faults, in *Proceedings of the Seventeenth International Conference on Architectural Support for Programming Languages and Operating Systems, ASPLOS XVII* (ACM, New York, 2012), pp. 123–134. https://doi.org/10.1145/2150976.2150990.

50. M. Hashimoto, W. Liao, S. Hirokawa, Soft error rate estimation with TCAD and machine learning, in *2017 International Conference on Simulation of Semiconductor Processes and Devices (SISPAD)* (2017), pp. 129–132. https://doi.org/10.23919/SISPAD.2017.8085281

51. J. Henkel, L. Bauer, N. Dutt, P. Gupta, S. Nassif, M. Shafique, M. Tahoori, N. Wehn, Reliable on-chip systems in the nano-era: lessons learnt and future trends, in *Proceedings of the 50th Annual Design Automation Conference, DAC '13* (ACM, New York, 2013), pp. 99:1–99:10. https://doi.org/10.1145/2463209.2488857.

52. J.L. Henning, SPEC CPU2006 benchmark descriptions. SIGARCH Comput. Archit. News **34**(4), 1–17 (2006). https://doi.org/10.1145/1186736.1186737

53. G. Hubert, L. Artola, Single-event transient modeling in a 65-nm bulk CMOS technology based on multi-physical approach and electrical simulations. IEEE Trans. Nucl. Sci. **60**(6), 4421–4429 (2013). https://doi.org/10.1109/TNS.2013.2287299

54. G. Hubert, L. Artola, D. Regis, Impact of scaling on the soft error sensitivity of bulk, FDSOI and FinFET technologies due to atmospheric radiation. Integr. VLSI J. **50**, 39–47 (2015). https://doi.org/10.1016/j.vlsi.2015.01.003

55. E. Ibe, H. Taniguchi, Y. Yahagi, K.I. Shimbo, T. Toba, Impact of scaling on neutron-induced soft error in SRAMs from a 250 nm to a 22 nm design rule. IEEE Trans. Electron. Devices **57**(7), 1527–1538 (2010). https://doi.org/10.1109/TED.2010.2047907

56. Imperas: Open virtual platforms (OVP) (2017). http://www.ovpworld.org/

57. Inside HPC: Supercomputers unlocking mysteries of the subatomic world (2018). https://insidehpc.com/2018/05/supercomputers-unlocking-mysteries-subatomic-world/

58. K. Johansson, M. Ohlsson, N. Olsson, J. Blomgren, P.U. Renberg, Neutron induced single-word multiple-bit upset in SRAM. IEEE Trans. Nucl. Sci. **46**(6), 1427–1433 (1999). https://doi.org/10.1109/23.819103

59. W.D. Jones, Building safer cars (2002). http://spectrum.ieee.org/transportation/advanced-cars/building-safer-cars

60. A.B. Kahng, The ITRS design technology and system drivers roadmap: process and status, in *2013 50th ACM/EDAC/IEEE Design Automation Conference (DAC)* (2013), pp. 1–6. https://doi.org/10.1145/2463209.2488776

61. M. Kaliorakis, S. Tselonis, A. Chatzidimitriou, N. Foutris, D. Gizopoulos, Differential fault injection on microarchitectural simulators, in *2015 IEEE International Symposium on Workload Characterization* (2015), pp. 172–182. https://doi.org/10.1109/IISWC.2015.28

62. D. Kanter, ARM goes 64-bit (2012). http://www.realworldtech.com/arm64/

63. T. Karnik, P. Hazucha, Characterization of soft errors caused by single event upsets in CMOS processes. IEEE Trans. Dependable Secure Comput. **1**(2), 128–143 (2004). https://doi.org/10.1109/TDSC.2004.14

64. M. Kooli, G. Di Natale, A survey on simulation-based fault injection tools for complex systems, in *2014 Nineth IEEE International Conference on Design Technology of Integrated Systems in Nanoscale Era (DTIS)* (2014), pp. 1–6. https://doi.org/10.1109/DTIS.2014.6850649

65. K. Kourou, T.P. Exarchos, K.P. Exarchos, M.V. Karamouzis, D.I. Fotiadis, Machine learning applications in cancer prognosis and prediction. Comput. Struct. Biotechnol. J. **13**, 8–17 (2015). https://doi.org/10.1016/j.csbj.2014.11.005

66. T. Li, J.A. Ambrose, R. Ragel, S. Parameswaran, Processor design for soft errors: challenges and state of the art. ACM Comput. Surv. **49**(3), 57:1–57:44 (2016). https://doi.org/10.1145/2996357

67. P.S. Magnusson, M. Christensson, J. Eskilson, D. Forsgren, G. Ha allberg, J. Hogberg, F. Larsson, A. Moestedt, B. Werner, Simics: A full system simulation platform. Computer **35**(2), 50–58 (2002). https://doi.org/10.1109/2.982916

68. M.M.K. Martin,, D.J. Sorin, B.M. Beckmann, M.R. Marty, M. Xu, A.R. Alameldeen, K.E. Moore, M.D. Hill, D.A. Wood, Multifacet's general execution-driven multiprocessor simulator (GEMS) toolset. SIGARCH Comput. Archit. News **33**(4), 92–99 (2005). https://doi.org/10.1145/1105734.1105747

69. T.C. May, M.H. Woods, A new physical mechanism for soft errors in dynamic memories, in *16th International Reliability Physics Symposium* (1978), pp. 33–40. https://doi.org/10.1109/IRPS.1978.362815

70. B. Mccluskey, Connected cars—the security challenge [connected cars cyber security]. Eng. Technol. **12**(2), 54–57 (2017). https://doi.org/10.1049/et.2017.0205

71. W. McKinney, pandas: a foundational python library for data analysis and statistics. Python for High Performance and Scientific Computing (2011). https://pandas.pydata.org/about/citing.html

72. S. Misera, H. Vierhaus, L. Breitenfeld, A. Sieber, A Mixed language fault simulation of VHDL and SystemC, in *Nineth EUROMICRO Conference on Digital System Design: Architectures, Methods and Tools, DSD 2006* (2006), pp. 275–279. https://doi.org/10.1109/DSD.2006.10

73. G.E. Moore, Cramming more components onto integrated circuits. Electronics **38**(8), 114–117 (1965). https://doi.org/10.1109/jproc.1998.658762

74. S. Mukherjee, *Architecture Design for Soft Errors* (Morgan Kaufmann Publishers, San Francisco, 2008)

75. S.S. Mukherjee, C. Weaver, J. Emer, S.K. Reinhardt, T. Austin, A systematic methodology to compute the architectural vulnerability factors for a high-performance microprocessor, in *Proceedings of the 36th Annual IEEE/ACM International Symposium on Microarchitecture, MICRO 36* (IEEE Computer Society, Washington, 2003), p. 29. http://dl.acm.org/citation.cfm?id=956417.956570

76. B. Nie, J. Xue, S. Gupta, C. Engelmann, E. Smirni, D. Tiwari, Characterizing temperature, power, and soft-error behaviors in data center systems: insights, challenges, and opportunities, in *2017 IEEE 25th International Symposium on Modeling, Analysis, and Simulation of Computer and Telecommunication Systems (MASCOTS)* (2017), pp. 22–31. https://doi.org/10.1109/MASCOTS.2017.12

77. A. Nordrum, Popular Internet of Things forecast of 50 billion devices by 2020 is outdated (2016). http://spectrum.ieee.org/tech-talk/telecom/internet/popular-internet-of-things-forecast-of-50-billion-devices-by-2020-is-outdated

78. E. Normand, J.L. Wert, H. Quinn, T.D. Fairbanks, S. Michalak, G. Grider, P. Iwanchuk, J. Morrison, S. Wender, S. Johnson, First record of single-event upset on ground, Cray-1 computer at Los Alamos in 1976. IEEE Trans. Nucl. Sci. **57**(6), 3114–3120 (2010). https://doi.org/10.1109/TNS.2010.2083687

79. J.M. Palau, G. Hubert, K. Coulie, B. Sagnes, M.C. Calvet, S. Fourtine, Device simulation study of the SEU sensitivity of SRAMs to internal ion tracks generated by nuclear reactions. IEEE Trans. Nucl. Sci. **48**(2), 225–231 (2001). https://doi.org/10.1109/23.915368

80. D. Pandini, Variability in advanced nanometer technologies: challenges and solutions, in *Integrated Circuit and System Design. Power and Timing Modeling, Optimization and Simulation* (Springer, Berlin, 2009), p. 2. https://doi.org/10.1007/978-3-642-11802-9_2

81. A. Patel, F. Afram, S. Chen, K. Ghose, MARSS: a full system simulator for multicore ×86 CPUs, in *Proceedings of the 48th Design Automation Conference, DAC '11* (ACM, New York, 2011), pp. 1050–1055. https://doi.org/10.1145/2024724.2024954.

82. F. Pedregosa, G. Varoquaux, A. Gramfort, V. Michel, B. Thirion, Grisel, O., M. Blondel, P. Prettenhofer, R. Weiss, V. Dubourg, J. Vanderplas, A. Passos, D. Cournapeau, M. Brucher, M. Perrot, D. Duchesnay, Scikit-learn: machine learning in python. J. Mach. Learn. Res. **12**, 2825–2830 (2011). http://dl.acm.org/citation.cfm?id=1953048.2078195

83. C. Perera, A. Zaslavsky, P. Christen, D. Georgakopoulos, Context aware computing for the Internet of Things: a survey. IEEE Commun. Surv. Tutorials **16**(1), 414–454 (2014). https://doi.org/10.1109/SURV.2013.042313.00197

84. P. Ramachandran, P. Kudva, J. Kellington, J. Schumann, P. Sanda, Statistical fault injection, in *IEEE International Conference on Dependable Systems and Networks with FTCS and DCC, DSN 2008* (2008), pp. 122–127. https://doi.org/10.1109/DSN.2008.4630080

85. S. Rehman, M. Shafique, F. Kriebel, J. Henkel, Reliable software for unreliable hardware: embedded code generation aiming at reliability, in *2011 Proceedings of the Ninth IEEE/ACM/IFIP International Conference on Hardware/Software Codesign and System Synthesis (CODES+ISSS)* (2011), pp. 237–246. https://doi.org/10.1145/2039370.2039408

86. S. Rigo, G. Araujo, M. Bartholomeu, R. Azevedo, ArchC: a systemC-based architecture description language, in *Sixteenth Symposium on Computer Architecture and High Performance Computing, 2004. SBAC-PAD 2004* (2004). https://doi.org/10.1109/SBAC-PAD.2004.8

87. F.R. Rosa, R.M. Brum, G. Wirth, F. Kastensmidt, L. Ost, R. Reis, Impact of dynamic voltage scaling and thermal factors on SRAM reliability. Microelectron. Reliab. **55**(9–10), 1486–1490 (2015). https://doi.org/10.1016/j.microrel.2015.07.013

88. F.R. Rosa, R.M. Brum, G. Wirth, L. Ost, R. Reis, Impact of dynamic voltage scaling and thermal factors on FinFET-based SRAM reliability, in *2015 IEEE International Conference on Electronics, Circuits, and Systems (ICECS)* (2015), pp. 137–140. https://doi.org/10.1109/ICECS.2015.7440268

89. O. Ruano, J. Maestro, P. Reyes, P. Reviriego, A simulation platform for the study of soft errors on signal processing circuits through software fault injection, in *IEEE International Symposium on Industrial Electronics, ISIE 2007* (2007), pp. 3316–3321. https://doi.org/10.1109/ISIE.2007.4375147

90. C. Russ, ESD issues in advanced CMOS bulk and FinFET technologies: processing, protection devices and circuit strategies. Microelectron. Reliab. **48**(8–9), 1403–1411 (2008). https://doi.org/10.1016/j.microrel.2008.07.042

91. A.L. Samuel, Some studies in machine learning using the game of checkers. IBM J. Res. Dev. **3**(3), 210–229 (1959). https://doi.org/10.1147/rd.33.0210

92. S.K. Sastry Hari, R. Venkatagiri, S.V. Adve, H. Naeimi, GangES: Gang error simulation for hardware resiliency evaluation, in *Proceeding of the 41st Annual International Symposium on Computer Architecuture, ISCA '14* (IEEE Press, Piscataway, 2014), pp. 61–72. http://dl.acm.org/citation.cfm?id=2665671.2665685

93. N. Seifert, Radiation-induced soft errors: a chip-level modeling perspective. Found. Trends Electron. Des. Autom. **4**(2–3), 99–221 (2010). https://doi.org/10.1561/1000000018.

94. N. Seifert, S. Jahinuzzaman, J. Velamala, R. Ascazubi, N. Patel, B. Gill, J. Basile, J. Hicks, Soft error rate improvements in 14-nm technology featuring second-generation 3D tri-gate transistors. IEEE Trans. Nucl. Sci. **62**(6), 2570–2577 (2015). https://doi.org/10.1109/TNS.2015.2495130

95. R. Shafik, P. Rosinger, B. Al-Hashimi, SystemC-based minimum intrusive fault injection technique with improved fault representation, in *Fourteenth IEEE International On-Line Testing Symposium, IOLTS '08* (2008), pp. 99–104. https://doi.org/10.1109/IOLTS.2008.25

96. Y.S. Shao, S.L. Xi, V. Srinivasan, G.Y. Wei, D. Brooks, Co-designing accelerators and SoC interfaces using gem5-Aladdin, in *2016 49th Annual IEEE/ACM International Symposium on Microarchitecture (MICRO)* (2016), pp. 1–12. https://doi.org/10.1109/MICRO.2016.7783751

97. P. Shivakumar, M. Kistler, S.W. Keckler, D. Burger, L. Alvisi, Modeling the effect of technology trends on the soft error rate of combinational logic, in *Proceedings of the 2002 International Conference on Dependable Systems and Networks, DSN '02* (IEEE Computer Society, Washington, 2002), pp. 389–398. http://dl.acm.org/citation.cfm?id=647883.738394

98. C. Slayman, Soft errors past history and recent discoveries, in *2010 IEEE International Integrated Reliability Workshop Final Report* (2010), pp. 25–30. https://doi.org/10.1109/IIRW.2010.5706479

99. M. Snir, R.W. Wisniewski, J.A. Abraham, S.V. Adve, S. Bagchi, P. Balaji, J. Belak, P. Bose, F. Cappello, B. Carlson, A.A. Chien, P. Coteus, N.A. Debardeleben, P.C. Diniz, C. Engelmann, M. Erez, S. Fazzari, A. Geist, R. Gupta, F. Johnson, S. Krishnamoorthy, S. Leyffer, D. Liberty, S. Mitra, T. Munson, R. Schreiber, J. Stearley, E.V. Hensbergen, Addressing failures in exascale computing. Int. J. High Perform. Comput. Appl. **28**(2), 129–173 (2014). https://doi.org/10.1177/1094342014522573

100. O. Subasi, S. Di, P. Balaprakash, O. Unsal, J. Labarta, A. Cristal, S. Krishnamoorthy, F. Cappello, MACORD: online adaptive machine learning framework for silent error detection, in *2017 IEEE International Conference on Cluster Computing (CLUSTER)* (2017), pp. 717–724. https://doi.org/10.1109/CLUSTER.2017.128

101. K. Tanikella, Y. Koy, R. Jeyapaul, K. Lee, A. Shrivastava, gemV: a validated toolset for the early exploration of system reliability, in *2016 IEEE 27th International Conference on Application-specific Systems, Architectures and Processors (ASAP)* (2016), pp. 159–163. https://doi.org/10.1109/ASAP.2016.7760786

102. J. Unpingco, *Python for Probability, Statistics, and Machine Learning* (Springer, Berlin, 2016). //www.springer.com/gp/book/9783319307152

103. M. Valderas, M. Garcia, R. Cardenal, C. Ongil, L. Entrena, Advanced simulation and emulation techniques for fault injection, in *IEEE International Symposium on Industrial Electronics, ISIE 2007* (2007), pp. 3339–3344. https://doi.org/10.1109/ISIE.2007.4375151

104. A. Vishnu, H.V. Dam, N.R. Tallent, D.J. Kerbyson, A. Hoisie, Fault modeling of extreme scale applications using machine learning, in *2016 IEEE International Parallel and Distributed Processing Symposium (IPDPS)* (2016), pp. 222–231. https://doi.org/10.1109/IPDPS.2016.111

105. L. Vlacic, M. Parent, F. Harashima, *Intelligent Vehicle Technologies: Theory and Applications* (Butterworth-Heinemann, Oxford, 2001). Google-Books-ID: oL6M0U3QEacC

106. J.T. Wallmark, S.M. Marcus, Maximum packing density and minimum size of semiconductor devices, in *1961 International Electron Devices Meeting*, vol. 7 (1961), pp. 34–34. https://doi.org/10.1109/IEDM.1961.187226

107. T. Watt, ARM discloses technical details of the next version of the-ARM (2011). https://www.arm.com/about/newsroom/arm-discloses-technical-details-of-the-next-version-of-the-arm-architecture.php

108. S. Woo, M. Ohara, E. Torrie, J. Singh, A. Gupta, The SPLASH-2 programs: characterization and methodological considerations, in *Proceedings of the 22nd Annual International Symposium on Computer Architecture* (1995), pp. 24–36

109. J. Yan, W. Zhang, Compiler-guided register reliability improvement against soft errors, in *Proceedings of the 5th ACM International Conference on Embedded Software, EMSOFT '05* (ACM, New York, 2005), pp. 203–209. https://doi.org/10.1145/1086228.1086266.

110. J. Yoshida, *Toyota Case: Single Bit Flip That Killed—EE Times* (2015). http://www.eetimes. com/document.asp?doc_id=1319903
111. M. Yourst, PTLsim: a cycle accurate full system ×86-64 microarchitectural simulator, in *IEEE International Symposium on Performance Analysis of Systems Software, 2007. ISPASS 2007* (2007), pp. 23–34. https://doi.org/10.1109/ISPASS.2007.363733
112. A. Zanella, N. Bui, A. Castellani, L. Vangelista, M. Zorzi, Internet of Things for smart cities. IEEE Internet Things J. **1**(1), 22–32 (2014). https://doi.org/10.1109/JIOT.2014.2306328
113. Y. Zhang, L. Peng, X. Fu, Y. Hu, Lighting the dark silicon by exploiting heterogeneity on future processors, in *2013 50th ACM/EDAC/IEEE Design Automation Conference (DAC)* (2013), pp. 1–7
114. J.F. Ziegler, W.A. Lanford, Effect of cosmic rays on computer memories. Science **206**(4420), 776–788 (1979). https://doi.org/10.1126/science.206.4420.776.

Index

Printed in the United States
by Baker & Taylor Publisher Services